COURAGE, ENDURANCE, SACRIFICE

The Lives and Faith of Three Generations of Missionaries

To Eddie
with love from
Charlotte

Charlotte Harris Rees

Copyright © 2016 by Charlotte Harris Rees

Courage, Endurance, Sacrifice: The Lives and Faith of Three Generations of Missionaries
Charlotte Harris Rees
www.AsiaticFathers.com
harrismaps@msn.com

Published 2016 by Torchflame Books
www.lightmessages.com
Durham, NC 27713 USA
SAN: 920-9298

Paperback ISBN: 978-1-61153-231-9
Ebook ISBN: 978-1-61153-230-2
Hardcover ISBN: 978-1-61153-235-7

ALL RIGHTS RESERVED
No part of this publication may be reproduced, stored in a retrieval system, or transmitted in any form or by any means, electronic, mechanical, photocopying, recording, scanning, or otherwise, except as permitted under Section 107 or 108 of the 1976 International Copyright Act, without the prior written permission of the publisher except in brief quotations embodied in critical articles and reviews.

What Others Say

Rich in historical facts, this biographical work of three generations of missionary work in one family covering the period from 1875 to 1972 and from "Wild West" Texas and Mexico then later to China and Taiwan is amazingly detailed. It vividly depicts the courage, endurance, and sacrifice of those pioneering missionary workers in faith.

In this well-researched book the author offers a good picture of the true nature of missionary work and of the contributions missionaries have made in many parts of the world.

For readers interested in the modern history of China during the most turbulent period of its history, this book also provides useful insights.

Dr. Hwa-Wei Lee,
Retired Chief of the Asian Division,
Library of Congress

In *Courage, Endurance, Sacrifice* Charlotte Harris Rees inspires readers in a carefully researched portrayal of three generations of her family as they faced epic degrees of adversity with simple, reasoned, obedient, and patient endurance. The family's selfless witness allowed them to become uniquely one with their indigenous people groups in a truly shared humanity. I recommend this page-turner.

Jim Berwick,
Archivist, International Mission Board,
Southern Baptist Convention

This book is dedicated to my daughter, Dawn Rees Powers.
Often you and I did not see eye to eye.
All along I loved you.
You have been God's beautiful gift to me.

Contents

What Others Say .... **iii**
Dedication .... **iv**
Introduction **1**
Main Characters in This Book 4
The First Generation: Missionaries to "Wild West" Texas and Mexico
William David Powell and Mary Florence Mayberry Powell **5**
Civil War Era – Murfreesboro, Tennessee 7
William's Education 8
Powell's Chapel Baptist Church 9
William Gets Married 11
William's Spiritual Mother – the Former Slave 11
Tuberculosis 12
W. D. Powell in Texas 12
John Westrup Martyred in Mexico 17
W. D. Powell and Family Go to Mexico 19
Saltillo, Mexico – the Powell's Original Mexican Home Base 20
Welcome to Saltillo 21
Madero Institute 23
Cooperation with Other Missionaries 24
Dangerous Trips 25
Inspiring Mexicans to Serve God 29
Health Problems of Early Missionaries to Mexico 31
The Powells Move to Toluca 33
Effective Ministry 34
The Conference on the Holy Spirit 35
Trials and Tribulations 36
Cuba 48
Back to Tennessee and Union University 48
Kentucky 50
Field Secretary for Foreign Missions Board 51
Who's Who in America 52
The Final Years 52

The Second Generation: Missionaries to China
Hendon Mason Harris Sr. and Florence Powell Harris 55
Early Missionary Work in China 57
Hendon and Florence Enter China 58
Kaifeng 60
Education of Women in China 63
The Role of Southern Baptist Missionaries to China 64
Life in Kaifeng 64
Evangelism in the Villages 67
The Worth of a Child 70
The Baptist Compound in Kaifeng 71
The Prayed for Son 73
Trip Home to America 74
Addie Cox 76
The Chinese Labor Corps in France 76
In Sickness and in Health 78
The Harris Family Returns to China 78
The Young Harris Children in China 79
Kaifeng Through a Child's Eyes 81
The Stursberg Compound in Kaifeng 85
Education of Foreign Children in China 86
Medical Maladies and Civil Unrest 87
Second Trip for Harris Family Back to America 88
Cows Travel to China 89
Hendon Harris Sr.'s Outstations 89
The Harris Family Returns to China Again 90
Japanese Occupation of Kaifeng 94
China Baptist Theological Seminary in Kaifeng 104
Missionary Families Ordered out of China 107
Back Home in Kaifeng 115
Hendon Sr. and Florence's Final Battle in China 117
Death of Son Eugene 118
The Final Years of Hendon Sr. and Florence 119
God's Work Goes On 120
The Third Generation: Missionaries to Taiwan and Hong Kong
Hendon Mason Harris Jr. and Marjorie Weaver Harris 123
Chinese Relationship to Native Americans 124
Hendon Jr. Runs Away 126
Hendon Jr.'s Education 126
Marjorie Weaver 127
Invitation to Go to Taiwan 135
Timeline 136

At Home in Taiwan (Also Known as Formosa) 145
The Orphanage at Banqiao 152
The Mountains of Taiwan 160
Che-ji-do (Jejudo) 167
Four of the Harris Children Leave Taiwan 172
Life in Indiana 174
The Harris Family Returns to Asia 179
My Dad Is Tops 185
The Divorce 190
The Map 191
Books by Dr. Hendon Harris Jr. 194
Addendum 1—Letter from Foreign Mission Board Rejecting Apostle Paul 195
Addendum 2—"The Prisoners of War Who Opposed the Chinese Communists—Never Return to the Mainland" 199
Addendum 3—"A Friend of China!—Dr. Hendon Harris [Jr.]" 201
Acknowledgements 205
About Charlotte Harris Rees 207
Notes 208
Additional Resources 223

Introduction

"Remember the days of old.
Consider the years of many generations."
Deuteronomy 32: 7

Several years ago I was asked to write a book about my father, Dr. Hendon Harris Jr. to explain the man who understood the strange Asian world map when he first viewed it.

This book answers that request. I have included my grandparents' and great grandparents' stories here because not only did they have exciting accounts to tell, but they also shaped my father's character. As it is said: "The apple does not fall far from the tree."

The three generations discussed in this text lived lives that were far from ordinary. They all also expressed interest in learning about from where Native Americans came.

This book has been years in the making. I wrote numerous pages in 2004 and 2005 but soon realized that if I were to tell about my father, I first had to explain his theories about ancient Chinese exploration in the Americas.

Therefore, I put this text aside but kept collecting information. In 2006 my abridgement of Father's nearly 800 page *The Asiatic Fathers of America* came out. That book was followed by three books of my own research which spring board from Father's and discuss different aspects of that subject. Today many around the world have embraced Father's interpretation of that map. Starting in 2016 a US based high school world history textbook, *Streams of Civilization*, Vol. 1, 3rd. ed., shows Harris's map and references his research.

Most families have legends. Mine has many stories – not all factual. However, in this book I tried to include only those that I could substantiate. Therefore, I have over 400 endnotes. Many of my endnotes are worth following up to the source so that one can glean even more of each account.

All three generations discussed here were prolific – including as writers of articles and letters. Each generation wrote at least one book. Therefore, I narrate as much as possible in each ancestor's own words.

My father's books and my parents' records and correspondence – which I collected over several years from various sources – were treasure troves for me. Mother left an audio recording of the story of her life.

How Beautiful the Feet, the autobiography of my grandmother, Florence Powell Harris, was my starting point for her generation and the one prior. Obviously I cannot retell every detail of her story or even narrate it as delightfully as she did. Therefore, I hope that readers will later go through her book, which my cousin Larry Harris has brought out again both in paperback and as an e-book.

Charlotte interviewing her Aunt Cita in 2015.

I interviewed several people who had known my father. Over the years I collected and read history books and numerous related missionary biographies. The excellent records of the Southern Baptists (including private missionary correspondence of these relatives—now open after being sealed for many years) and reports of other mission agencies furnished insights and documentation.

What I read in the research for this book played a symphony on every string of my emotions, but the journey was adventuresome. I hope like I did—that perhaps you will laugh, cry, get angry and/or happy, but most of all be inspired.

In celebration of my father's life this book is being published in 2016 – the year he would have turned 100. 2016 is also the year that my sweet Aunt Cita turns 90. She is our only relative remaining from her generation.

None of us choose our forefathers, nor can we claim credit or blame for what they did or did not do. However, in understanding our ancestors and their worlds perhaps we can better understand our own.

I love hearing from readers. E-mail me at HarrisMaps@msn.com and/or review this book on Amazon and other sites.

Charlotte in British Columbia in 2010 interviewing Peter Stursberg, who in Kaifeng, China, had been a childhood friend of her father, Hendon Harris Jr. As an adult Stursberg was a WWII Correspondent in Europe and was well known as an author of several books.

Main Characters in This Book

Several of the people in this text have similar names. Below in alphabetical order of given names is a list of the main characters. In case one becomes confused about whom I am discussing, he can refer to this list below.

- **Aurora Dawn** - Daughter of Hendon Jr. and Marjorie Harris
- **Charlotte** - Daughter of Hendon Jr. and Marjorie Harris
- **Cita (Florencita) Harris Strunk** - Daughter of Hendon Sr. and Florence
- **Florence Mayberry Powell** - Missionary to Mexico, wife of William Powell
- **Florence Powell Harris** - Missionary to China, daughter of the Powells
- **Hendon Harris Jr.** - Missionary to Taiwan and Hong Kong, husband of Marjorie
- **Hendon (H. M.) Harris Sr.** - Missionary to China, spouse of Florence P. Harris
- **Hendon III** - Son of Hendon Jr. and Marjorie Harris
- **John** - Son of Hendon Jr. and Marjorie Harris
- **Lillian** - Daughter of Hendon Jr. and Marjorie Harris
- **Marjorie Florence** - daughter of Hendon Jr. and Marjorie Harris
- **Marjorie Weaver Harris** - Missionary Taiwan and Hong Kong, wife of Hendon Jr.
- **Mejchahl** - Daughter of Hendon Jr. and Marjorie Harris
- **William (Willie or W. D.) Powell** - Missionary to Mexico, dad of Florence Harris

The First Generation: Missionaries to "Wild West" Texas and Mexico

William David Powell and Mary Florence Mayberry Powell

W. D. Powell.

During the United States Civil War, my great great grandmother, Nancy Rankin Powell, was dying. Realizing that slaves would soon be freed, she turned to her servant – a sister in faith – and asked a seemingly impossible favor. Nancy begged that after emancipation the slave stay on with the family until Nancy's young son, Willie, made a personal decision to place his faith in Christ.

Although the faithful ex-slave did stay until Willie left home, Willie did not make a profession of faith until later. Meanwhile, the slave's fervent prayers continued going up for him.

I am not proud of the fact that my relatives owned slaves. However, I am grateful that the prayers and actions of this humble lady helped lead to three generations of Baptist missionaries to foreign countries plus numerous other full time Christian workers. This is not a story of sinless saints but rather of ordinary people who trusted in God. It is a story of faith, forgiveness, love, and sacrifice – answers to the humble slave's prayers.

My great grandfather, William David Powell, was known as Willie during his childhood, then William or W. D. Powell, later in life. He was born July 1, 1854 in Madison, Mississippi to Nancy and William Madison Powell – known by his friends as Madison.

Madison Powell in later years.

The story is told that shortly after William's birth, because of his mother's poor health, the family quickly moved back to Murfreesboro, Tennessee where most of their relatives lived. On their arrival home, Nancy's mother gave them slave Laura Ward as a wet nurse for Willie.[1]

Laura Ward was the family slave. However, it is uncertain whether she served as a wet nurse in Tennessee. My cousin, Larry Harris, found whom he believes to be Willie and his parents on the 1860 census – still in Mississippi. We are certain that Willie was born in Mississippi in 1854 and that sometime before his mother's death in 1862 the family moved back to a farm near Murfreesboro, Tennessee.

Civil War Era - Murfreesboro, Tennessee

Nancy Powell died during the US Civil War between the Union and Confederate States. The Emancipation Proclamation, which freed slaves, was issued on January 1, 1863 – about 6 months after Nancy's death.

The Civil War battles in Murfreesboro were not far from the Powell home. Years later William recalled walking over those battlefields as a young child and viewing the dead and wounded.[2] The occupying Union troops would not allow Nancy's body to be carried through their lines to the local cemetery. Consequently, she was buried on the Powell family farm.[3]

Invading Union soldiers stole all available food from surrounding farms. The Powell family, along with their neighbors, were in dire straits. Then the Union soldiers came back and took the Powell's last milk cow.

Madison decided to send Willie, accompanied by the servant, to Union Army headquarters to ask for the return of the cow. The 8-year-old boy was allowed to speak to the commanding officer, who asked Willie what he wanted. Florence Powell Harris (my grandmother) wrote:

> Willie burst into tears and between sobs sputtered out, "Your soldiers have taken our last cow, and we have nothing left to eat." The officer was moved. He commanded…[his soldiers]…to see to it that [the cow] was returned to him.[4]

Before they were robbed of their goods, Willie had been carrying buttermilk and vegetables to both Union and Confederate hospitals.[5] Secretly, under patches sewn onto his clothes by the slave Laura Ward, he had also been taking messages across Union lines to wounded and captured Confederate soldiers. Unsuspecting of Willie's secret mission, but apparently cognizant that he frequently visited Confederates (known as Rebels), Union troops called him "Rebel Willie."[6]

Within a few years Willie's father, remarried. His second wife was Jane Mayberry and they bore a son named Joseph Daniel Powell. Around 1867 Jane's eight year old niece was orphaned so moved into the Powell home.[7]

At that time living conditions were still dire because of after effects of the Civil War. Florence Harris wrote that when Willie was 13 years old, his father handed him an ax and said: "Willie,

I am giving you this ax," and pointing to the woods, [on his 160 acre farm] added, "There are the trees.... From now on, you are on your own."[8]

Florence Harris wrote that William was able to support himself and his education by selling stove wood but at times was so poor that his shoes had to be strapped to his feet.[9]

William's Education

So Great a Cloud of Witnesses, a book about the history of Union University, states that in 1871 William Powell graduated from Union, then located at Murfreesboro, Tennessee, with a Master of Arts Degree. William was only 16 years old then!

At that time Union accepted students as young as 14 years old. There were numerous required classes and recitations but the number of years to graduation was "dependent upon the capacity and application of the student" in passing proficiency exams in various subjects. The highest degree that Union offered at that time was Master of Arts. *So Great a Cloud of Witnesses* lists William among Union University's distinguished alumni[10] and their online time line gives his name and graduation date.

Union University Murphreesboro.

William wrote about his college years:

> To meet the expenses incident to my college life, I hauled two loads of wood to Murfreesboro every Saturday. I usually cleared five dollars thereby which was sufficient to pay for my books and tuition. To further increase my income I also bought and sold sheep and cattle. I began this at fourteen years of age.[11]

At that time schools of higher learning were separated by gender. William, in the tradition of Union's then all male student body, most likely lived at home with his parents. The timing of William's education was fortunate. Union at Murfreesboro closed its doors in 1873 – two years after William graduated[12] – when this institution joined with other universities and moved to Jackson, Tennessee.[13]

A historical marker states that Union University was charted in 1848 by the Baptist General Association of Tennessee. It was closed during the Civil War when it was used by both armies as a hospital. The marker states that Union reopened in 1868 but in 1873, due to a cholera epidemic, moved to Jackson.

It is said that a man is known by the friends he keeps. One of William's friends and classmates at Union, George M. Savage, five years older than William, later became President of Union University.[14] Their friendship continued for the rest of their lives.

Sometime during 1871, the year William graduated from college, he placed his faith in Christ as his personal Savior. He then decided to go to seminary to study for the ministry.[15] Surely this was an answer to the slave's prayers.

William stated:

> My seminary training was received in the Southern Baptist Theological Seminary [SBTS] which at that time was located at Greenville, S. C. At the age of seventeen I taught school to meet the expenses of my theological training.[16]

On January 1, 1874 William was ordained to the Baptist ministry.[17] Also that year, just before his 20th birthday, William had earned a master's degree at SBTS.[18] Returning to Murfreesboro he continued as a school teacher, then in 1875 was president of Oakdale Academy[19] and also preached at Ellis Chapel (later called Florence Baptist Church).[20] Later Baylor University, Texas, awarded him a third master's degree – that one honorary.[21] William received an honorary Doctor of Divinity degree from Union University in 1884.

Powell's Chapel Baptist Church

In the summer of 1875, 21-year-old William on his way home, rode past an abandoned Methodist church in Murfreesboro. He wondered whether he could start a Baptist church there, so held a series of meetings in that forsaken building. Bethany Hawkins stated in her 2014 master's thesis on the history of Powell's Chapel Baptist Church:

> At the final meeting on July 2….Powell as moderator, called for a ballot to name the church, and the members unanimously chose Powell's Chapel Baptist Church…. Powell did not agree with their choice. In a 1930 letter detailing the founding of the congregation William wrote, "When it was proposed to call it Powell's Chapel I begged them not to do it, but they did."[22]

A problem ensued. Simpson Harris, the husband of one of the founding members of Powell's Chapel, was a member of another denomination. Hoping that his wife would worship with him rather than at Powell's Chapel, Harris tried to undermine the upstart church. Hawkins wrote:

> Shortly after the first...meeting, Harris arrived at the church...during a worship service with wagons and servants. He had purchased the building and planned to remove it to his property....the congregation sang the last hymn and then [Harris] began dismantling the structure.[23]

Hawkins continued by quoting a letter written by William Powell:

> "The wagons were immediately placed in position and ladders set against the walls. The men scaled the ladders and began to remove the roof, while many of the worshipers looked on.
>
> ".... [Harris] had his negroes to tear it down and move it to his home and made a barn of it and filled it with fodder. The Lord God struck his house with lightening and burned it and nobody ever heard of my [Powell] shedding any tears."[24]

Nevertheless, the group built a new meeting house. According to William, Joe Putnam, a friend of his father even offered to build one long wall of the church.[25]

However, soon William faced a major health problem and his physician advised him to move to Texas. After only three months as pastor, William turned the reins over to the next parson.

Today Powell's Chapel Baptist Church, in Murfreesboro, Tennessee, a church named after 21-year-old William Powell, is still a vibrant congregation, and retains the same name.[26]

Photo courtesy of Bethany Hawkins.

William Gets Married

The Descendants Database of my membership with Daughters of the American Revolution in Washington, DC, shows that William Powell married Mary Florence Mayberry on October 5, 1875 in Rutherford County, Tennessee.[27] Florence, as she was known, was then only 16 years old and William was 21.

Florence Mayberry (variously spelled Maberry or Mabry) was the orphaned niece of William's step-mother. Though no relationship to William, by then she had lived in the Powell home for eight years after her parents died.[28] Because of their age difference, William had paid little attention to Florence – until another young man asked William whether Florence was seeing anyone. On the spot William decided to claim Florence as his own. Shortly after that they married.

William's Spiritual Mother – the Former Slave

William considered Laura Ward, the slave (by then freed), to be his spiritual mother and called her Aunt Laura.[29] In later years, whenever William returned home to Murfreesboro, he was frequently asked to preach at various churches.

Florence Powell Harris (William's daughter) wrote:

> When he returned [to Murfreesboro], his friends, knowing his affection for his old colored Mammy [apparently Florence did not know Laura's name], would see that she was suitably dressed, even starching her bonnet, for she was quite feeble, and escorting her to church, made sure she got a front seat where she could see and hear her Willie preach.
>
> ...when Mammy [Aunt Laura Ward][30] died, Father assumed her burial expenses. She was laid to rest beside his own mother's grave, for Mammy was his spiritual mother.[31]

In that era, people of color were not normally allowed to sit in the front of white churches nor were they buried next to white people. William did not give in to protocol of that day. Though churches in that era were segregated, all of his adult life William had special affection for black congregations.[32] He believed that in God's eyes all people are equal.

We have not been able to establish whether Laura Ward was Negro or Cherokee [Native American] or to find out any more details about her. Ward was well known as a Cherokee surname in that time and location. In that era some Cherokees were slave owners, while others were slaves who took on the surname of their masters.

According to a report on the website of Luginbuel Funeral Home, Laura Ward was buried on the old Powell property to the left of Willie's mother, Nancy. That record states that Willie paid for a headstone for Aunt Laura, but it was later removed for unknown reasons.[33]

Tuberculosis

According to a letter written by W. D. Powell in 1930, he moved to Texas in 1875 under orders from his doctor, who had diagnosed him with "tuberculosis in an incipient state [beginning stages]."[34] Apparently it was believed that the drier climate of Texas would be helpful.

X-rays did not begin until 1895 – twenty years after William's diagnosis.[35] In William's era most people did not understand that tuberculosis (TB) was contagious. Effective treatment was not available until the mid-20th century.[36]

Tuberculosis is an ancient disease sometimes referred to as consumption or "white plague (because of the extreme pallor [paleness] of those infected)."[37] In the late 19th century tuberculosis was one of the major causes of death in the Western world.[38] In one 1916 study 50 percent of the participants were dead within five years. This diagnosis in the year that they married had to have been a terrible blow to the young Powell couple.

W. D. Powell in Texas

In *West Texas Historical Association Year Book,* October 1933[39] William related that in 1875 he and his wife Florence took a train to Texas where a church in Ft. Worth had called him to be pastor. However, he never got to that church. At that time Ft. Worth was very small and did not have railroad connections. To get there William and Florence would have had to have taken a stage coach from Dallas.

Some other pastors that William met on the train to Texas told him that they were sure that Ft. Worth would never amount to anything. Those preachers got off the train at Mineola, Texas for some services.

In Dallas a telegram caught up with William inviting him to pastor a church in Mineola. Leaving his wife in Dallas, William returned to Mineola to scope out the situation.

The very small town of Mineola wanted William to pastor the church gratis. The attractive income offered was to come from teaching at the local school. That institution had been closed for two years after the town bad boys repeatedly misbehaved and almost gouged out the eye of the former teacher.

Unlike Ft. Worth, which has grown to be a major metropolis, Mineola was and still is a small town. When William arrived he discovered that it was what he described as "a one man town."

Dr. Scruggs, the man behind the generous dual job offer, was the town physician, pharmacist, postmaster, traveling preacher, real estate developer, music instructor, insurance salesman, farmer, brick mason, inventor, and was part owner of the newspaper, a dairy, butcher shop, jewelry store, boarding house, and lumber business. William stated that in all Dr. Scruggs had 23 jobs.[40] William wrote:

> My health was poor and I could not get insurance in an old line company. He [Dr. Scruggs] was the agent of a wild cat company at Dallas and sold me a policy for sixteen hundred dollars without charging an agent's fee.[41]

Map by Dave Rees.

Although sounding like a preposterous amount for health insurance during that era, William was quite ill and desperate. In other ways Scruggs was very generous to William – giving him the lumber to build his house and making sure that he was able to purchase the land for the house at half price.

William served as principal at Mineola High School from late 1875 through spring 1877. Though there were some trying times, he was able to control the students. William Eaton Powell, the Powell's first son, was born July 1876, in Mineola.

Due to his lung condition William needed the drier air of West Texas - generated by the Chihauhuan Desert. Therefore, after two school years in Mineola, William and his family headed to San Marcos, a frontier town, to take a pastorate that had been offered to him. Again, he never arrived. On the way to San Marcos William attended the Baptist State Convention where he was elected Sunday School evangelist.

It must have been an act of God to turn sickly William into the cyclone that happened in Texas in the next five years. Today he is mentioned in almost every book on the history of Baptists in Texas.

From 1877 to 1882 William served as Secretary of the Sunday School Convention of Texas Baptists.[42] During that time he started 400 Sunday Schools involving 20,000 people[43] across the height and breadth of that huge state. Many of those Sunday Schools were in places which prior to that time had no religious work.[44] From those Sunday Schools Baptist churches sprung up across Texas.

William wrote that when he first went to Texas the state had only 1400 miles of railroad so most of his travel was by horse. He travelled first on a dapple gray pony then in an open top buggy. Later he graduated to a hack pulled by two horses. William described the hack as:

> ...a four-wheel conveyance large enough to hold a considerable quantity of supplies, books, and bedding. This was necessary particularly during the last years of my work when frequently I would be obliged to camp far away from any human habitation. I always carried a rifle both for protection and for hunting. Fortunately I never had to use it except to secure wild game for my own sustenance.[45]

By that era most of the wild bison that once roamed Texas had been killed. However, Native Americans, also known as Indians, still wandered the plains.

A 1930 article by I. G. Murray discussed William's work with Sunday Schools in Texas. Murray quoted William:

> "On a Sunday afternoon, forty miles from San Antonio, the first organization was effected. The following Lord's day the wild Indians came and stampeded the horses and drove them off. A posse of men pursued them and engaged them in battle at Del Rio, slaying one Indian and recovering the horses.
>
> "Near my home in Burnett [Burnet] County were sixteen graves of people, fifteen of whom had been slain by the Indians. Their war whoops were frequent. They made their raids every full moon."[46]

The Sunday School movement started in the late 18th century. *Christianity Today* magazine states:

> Sunday schools were originally... places where poor children could learn to read...By the mid-19th century, Sunday school attendance [in the USA] was a near universal aspect of childhood. Even parents who did not regularly attend church themselves generally insisted that their children go.... Religious education was...a core component. The Bible was the textbook used for learning to read....Inculcating Christian morality and virtues was another goal of the movement.[47]

About the time William moved to Texas, public education became mandatory. *Christianity Today* indicates that after that mandate Sunday schools became strictly religious. That article states: "Nevertheless, many parents continued to believe that regular Sunday school attendance was an essential component of childhood."[48]

History of Texas Baptists (1900) by Benjamin Fuller outlined William's yearly progress in working with the Sunday Schools. In his first year 95 Sunday Schools were organized, 3800 people joined the new schools, and 2800 added to existing schools. The second year there were 88 new Sunday Schools and William traveled 6000 miles, mainly on horseback.

Many of the settlers in Texas welcomed William's work there – however, not everyone. Those were the days of the Wild West and William met numerous unsavory characters including horse thieves and murderers. At one location in the middle of a sermon William discontinued an illustration he was giving. Another minister had tugged William's coattail to alert him that an insolent known murderer in the audience was squirming and fidgeting with his pistol.

At another place William met a man who had left to get firewood but on returning home found his wife and young children had been scalped by Indians. That man showed William the 11 Indian scalps he had taken in revenge.

In one group of 14 men to whom William preached near the Mexican border, 13 of them were college graduates. They had gone to that desolate area for health, adventure, or to make money in sheep or cattle. William stated:

> These men were all "batching" and lived in dugouts or one room box cabins. Their system of keeping house was similar in all cases. They would cook corn bread in a skillet. Their loaves were thick…and soggy and would generally sour before all had been eaten; but soured or otherwise we ate it and thrived. This bread with bacon and coffee and occasionally beans constituted the standard fare of pioneer stockmen.[49]

In one place William recommended a local resident as teacher for the Bible class. However, the community rejected that man as too worldly. The only charge against the man was that he had ordered a wood floor for his house when everyone else there had dirt floors.

In addition to shady characters William was in danger both from the weather and wild animals. Coyotes were rampant and would steal food. Once in order to protect his provisions William slept with his bacon under his saddle, which he used as his pillow. Nevertheless, a coyote stole the bacon while he slept.

On another occasion William and another man were almost run over during blizzard conditions when a thunderous stampede of thousands of cattle ensued – lasting most of the night. Although they had a fire, William and his friend almost froze. The next day not far away they came upon approximately 250 sheep in a huddle that had frozen to death.[50]

William reported seeing what surely must have been the famed Buffalo Soldiers. He stated:

> At San Angelo I preached the first sermon ever preached in the community by a Baptist. Later I carried [seminary graduate] Edwin Mayes there and organized a church and preached there on several other occasions. There were very few white people in the community at the time and most of these had little interest in religion.
>
> The population was made up principally of negro soldiers and their camp followers, a number of Mexicans, a few traders and merchants and their families and some cattle and sheep ranchmen. Conditions were such that we did not dare to have services at night. A famous Methodist preacher named A. J. Potter preached occasionally at San Angelo during these years. He always had a gun with him when he went into the pulpit. I always left my gun outside.[51]

The Ninth United States Cavalry, composed mainly of black soldiers, was sent to West Texas after the Civil War to protect settlers from Indian attacks. That was the first time that black soldiers served in a US peacetime army.

In a previous book I discussed the Buffalo Soldiers in West Texas:

> The Indians nicknamed them "Buffalo Soldiers" because of their dark skin, curly hair, and fierce fighting spirit. The Buffalo Soldiers adopted that name with pride and even put a buffalo on their regimental crest.[52]

Part of William's job in Texas involved selling Bibles and religious materials. Because he had sold more than any other representative, he was invited to address the Northern Baptist Societies in New York in 1881. William wrote that he was so deeply tanned at that time from his travels that a news reporter in New York referred to him as: "a distinguished colored minister from Texas."[53]

Several people to whom William ministered during those years later became preachers or generous donors to missions. William felt that those years in Texas were among the best of his ministry.

In his resignation letter at the end of five years William stated:

> Nearly 5 years ago in this house I was requested to become your missionary. After prayerful consideration I accepted.
>
> Through heat and cold, dust and rain, hardship and discouragement, success and victories I have tried to cling to Christ and go on with the work. Let us praise God for whatever of good has been accomplished during the past 5 years.
>
> What change God hath wrought in our Texas Zion within these 5 years. Our churches in a large measure have thrown off the lukewarmness and

> indifference once so prevalent. Called to work in the Sunday Schools it has increased their piety and their contributions to home and foreign missions.[54]

William resigned from his position in Texas in order to go to missionary work in Mexico. In that last year in Texas he delivered 299 sermons and addresses, conducted 45 prayer meetings, visited 488 families, saw 145 conversions, and travelled 9500 miles.[55]

John Westrup Martyred in Mexico

William met Thomas Westrup at a church conference in Texas. Thomas related that his brother John, a fellow Baptist, was evangelizing in Mexico so William helped raise funds from the Texas Baptists for John's work.

After evangelizing in Mexico less than a year, John Westrup had already started four churches[56] when he was murdered and his mutilated body thrown into a cactus patch.

The 1880 Southern Baptist International Missions Report for the Mexican Mission stated:

> Some are of the opinion that the Mexicans, disguised as Indians, had known Brother John in Santa Rosa de Musquiz as a Protestant preacher, and butchered his body so, through hatred to Protestants.[57]

According to *Baptists and Mission:* "Mexicans in that day did not see fit to investigate the murder of Protestants."[58] About 20 years earlier – around 1861 – religious toleration had begun in Mexico for non-Catholics, so Protestants started to enter that country.[59] However, old traditions die hard.

William Powell was sent by the Texas Baptists to investigate Westrup's death. Since Texas, where William was ministering, borders Coahuila, the Mexican state where Westrup was killed, William was the logical choice to go. *An Evangelical Saga* states: "He [William] was deeply moved by reading the blood-stained diary [of John Westrup] found at the scene of the crime."[60]

In *Heroes and Martyrs* (1911) William was quoted:

> Word came to us in December that John O. Westrup had been murdered. At the suggestion of brethren in Texas I hired two men and went to ascertain the true facts, and my father took his trusty rifle and went with us. It was a long, tedious, dangerous trip after we passed beyond the suburbs of San Antonio. One of us had always to be on guard at night while the other three slept. There were but few houses from San Antonio to Ladero [sic – Laredo], an insignificant town without any railway connection.
>
> We had stirring adventures in Mexico. Several nights we did not sleep because we could see men prowling around our camp. One night four

> Americans camped in sight of us and neglected to stand guard, and the next morning two were dead and another was dying.
>
> In Lampazos, Mexico, we found many of those baptized by Brother Westrup. Some of them said, "God has sent you in answer to our prayers for someone to take up Brother Westrup's work."
>
> I went to the scene of the murder, looked on Westrup's new grave and prayed the Lord to give us Mexico for Christ. I found that one Saturday Brother Westrup had been overtaken while partaking of his noon-day lunch on the road to Mosquiz [Muzquiz] near Progresso.
>
> He was killed and stripped of his clothing and his body was mutilated and pitched upon a "Spanish dagger" (a yucca palm tree) and left there. A Mexican brother who accompanied him was also killed. I found a piece of his day book [diary], stained with his own blood, telling of at least seventy-five people who had been converted and baptized.
>
> I was profoundly impressed with the zeal and Scripture knowledge of the young converts. Several of them could quote chapter after chapter of the Old and New Testaments. All of them were witnesses for Christ.
>
> I visited Westrup's widow and his orphan children [then rendered financial aid to them].[61]

Following that trip to investigate the death of Westrup, William felt the call of God to go as a missionary to Mexico but hesitated because at that time he had a wife and three children – William age 6, Anita 4, and Nell 2. Neither did he see how he could leave his work in Texas.

The Southern Baptist Convention based in Richmond, Virginia, already had one missionary in Mexico and decided to explore the feasibility of expanding their work into the Mexican state of Coahuila. Justice Anderson wrote in "Nineteenth-Century Baptist Missions in Mexico:"

> In 1882, General A. T. Hawthorne, a retired Confederate officer who was an agent of the FMBSBC [Foreign Missions Board of the Southern Baptist Convention] sent Powell and O. C. Pope, an official of the ABHMS [American Baptist Home Missions Society], to Mexico to reconnoiter the northwest area as a possible mission field.
>
> Accompanied by [William] Flournoy [who had just started working in Progresso, Mexico – state of Yucatan – under the auspices of the Southern Baptists],[62] they explored the whole rugged area of Coahuila in a horse-drawn buggy. On their return, they gave a sanguine report that caused the FMBSBC to establish Mexico as its official mission field.[63]

In March 1882 General Hawthorne penned a four page tightly spaced letter to Dr. Tupper of the Southern Baptists in Richmond praising William and strongly recommending that they send William to Mexico but stating that if they did not, William would probably go there under the Northern Baptists. In that letter Hawthorne wrote regarding W. D. Powell:

> He is universally beloved of the entire Denomination, and there is no man in the State, for whom Support they would more heartily unite...His name is as familiar as household words and it really seems as if every woman and child in Texas loved him.....I know of few or any that I would select in preference to him....this may sound...extravagant, but if you should take the trouble to investigate the matter, you would find that I have not exaggerated...[64]

Hawthorne's letter also made it clear that when Powell went to Mexico he planned to go with his wife and sister-in-law – they were a package. The three were appointed that May.

W. D. Powell and Family Go to Mexico

That same year William Powell's wife, their three children, and his wife's sister, Annie Mayberry accompanied him to Mexico.

At that time William and Florence owned houses in two locations and in a third place an acreage on which they had cattle. They left their holdings in the care of a trusted Baptist judge with instructions to sell their assets and to place the proceeds in the bank.[65]

William stated:

> ...the outdoor life [in Texas] improved my health appreciably. If I had stayed in eastern Texas I probably should not have lived to be forty. The balmy air of the frontier country [West Texas] gave me some relief and shortly after I went to Mexico in 1882 the tuberculosis left me as completely as a snake's old skin leaves its body in spring.[66]

William wrote:

> I was appointed as missionary to Mexico by the Foreign Mission Board of the Southern Baptist Convention [in 1882].... The Mexican government compelled me to carry a Winchester rifle, and furnished me an armed escort when I ventured beyond the city limits....I travelled thousands of miles on horseback...[and] baptized twelve hundred and organized thirty-one churches.[67]
>
> ....it was decided that I should locate at Saltillo...Leaving my family, I went to Saltillo, rented a house and secured it by law for three years. It was well known that so soon as people ascertained that I was a Baptist, much less missionary, I could not have rented a house in that city of twenty thousand people.

> The night I returned to Monterrey, two of our children [became ill with] yellow fever. We were stopping in the home of the brother of the murdered missionary [Westrup].
>
> There were 34 members in the Monterrey Baptist Church, and the next day all of them quit work and came to show sympathy in our distress... The second night [the physician] told me that he did not believe that the children would live until morning.
>
> Soon I noticed that all the Mexicans had left the room and house. Brother Westrup said...they have gone to our little chapel to stay on their knees praying until the children are dead or better. The children were spared.[68]

Saltillo, Mexico – the Powell's Original Mexican Home Base

Thirty Years in Mexico states:

> W. D. Powell, his wife and her sister, Miss Anna J. Mayberry, appointed by our Board in May, reached Saltillo, October, '82.
>
> Only he who has experienced it can realize the strange feeling that creeps over the foreign missionary on reaching his field. The country, climate, people, customs, houses, food, clothing, language, religion – everything entirely different from what he had before known. He is "slow of speech, dull of hearing" and as helpless and a little child. He feels like he has awakened on another planet.[69]

William wrote:

> I studied Spanish at the state college in the forenoon and gave lessons in English to the Mexican people in the afternoon. I sought diligently to obtain a speaking knowledge of the Castillian [sic] tongue. One morning I met the mayor...and he said to me in Spanish: "Good morning! How are you? How is your family, and how is your wife?"...I weighed all my words and my reply, instead of saying that my wife had a slight headache, I used a similar word but with entirely different meaning which made me say that she was drunk from drinking too much beer. Ever afterward when we met he would inquire if my wife had sobered up yet.[70]

However, William took Spanish lessons only a short time before he preached in that language. He stated:

> Finding preaching through an interpreter so soulless, I dismissed him, and in four months the Lord enabled me to preach so that the people could understand.[71] [Perhaps they also had a few chuckles!]

William accepted a position teaching English at the college in Saltillo in order to make inroads with that sector of society.

Florence and her sister both struggled to learn Spanish. Florence wrote:

> My sister's [Annie Mayberry] school is doing quite as well as we could expect, considering her limited knowledge of the language. She and I having never studied any foreign language, find it more difficult to acquire the Spanish than my husband.[72]

From his first year in Mexico William taught young Mexicans in leadership. One Mexican man rode 70 miles on horseback to inquire about training with William. Meanwhile, William also pled for American missionaries to come help.

Frank Patterson stated in *A Century of Baptist Work in Mexico:*

> Powell was a man of action. He had been in Saltillo only three months, when he organized the First Baptist church of Saltillo with eight members on January 28, 1883.
>
> ....Powell could not confine himself to pastoring a church. A few months after organizing the church in Saltillo, he turned the pastorate to Porfirio Rodriguez.[73]

At that time William began traveling on evangelistic trips. William credited Alejandro Trevino, brother-in-law of the Westrup brothers, and Porfirio Rodriguez, who had also co-labored with the martyr, with helping establish the work in Saltillo.[74] Trevino had graduated from Monterrey Normal College. Rodriguez, also a native Mexican, had been educated at Southern Baptist Seminary in Louisville.

Welcome to Saltillo

William, the eternal optimist, in his initial mission report praised the climate of Saltillo, their new hometown. He described Saltillo, at 4500 feet elevation, as "perpetual spring" – never lower than 52 or higher than 82 degrees Fahrenheit.[75] Though the temperature was ideal, their initial reception was not.

William's first mission report to the Southern Baptists stated:

> While we were at worship on Sunday morning, some of the devil's emissaries came to the window and spit on us. At night the crowd was very boisterous and I thought they would have stoned us.
>
> When I asked persons who wished to join the church to come forward, one that spit on us approached to bring on some trouble. I talked so kindly about the necessity of saving his soul that he burst into tears, and the mob dispersed.

> On Thursday the attendance was larger than usual, which incensed the crowd and they stoned the house. Some thought they intended to break in and kill us. The Mayor provided police protection.[76]

Map by Dave Rees.

In that same mission report William's wife, Florence, added:

> ....I was much alarmed the night we were stoned. And I am sometimes apprehensive lest...some wicked person...injure my husband. We are sincerely hoping for reinforcements soon.[77]

William's daughter, Florence Powell Harris, wrote:

> Twice in the early days of Father's ministry in Mexico, the Powell family faced annihilation at the hands of national religious fanatics. But God intervened.[78]

On one of those two occasions the local American physician, Dr. Bibb, faced off the angry, howling crowd gathered at the Powell's gate[79] until they left one by one.

At another time a man came to one of Powell's meetings with the intention of killing him. William stated:

> Juan Chavez was a fanatical opposer of our missionary work and came with some rowdies to murder me, but he heard the Gospel, was converted and baptized and has been my life-long friend. For five years he traveled with me over the mountains of Mexico....When I was in imminent peril, he never failed to place himself between me and danger.[80]

Another book quoted William describing an incident involving Juan Chavez:

> At Rayones, as I was about to baptize a young man, his brother rushed up saying that if the brother was baptized, he would kill us both. I asked the candidate what he wanted to do. He replied that he wanted to obey Christ. I started down into the water and the infuriated brother started to make good his threat, when others disarmed him.
>
> That night he came to the services to shoot me, but Brother Chavez stood before me to shield me while I preached. The Lord convicted him [the angry brother] of sin before we concluded the services and he hid his pistol, begged my forgiveness and offered himself for baptism. He died four weeks later, happy in the Savior's love.[81]

A Century of Baptist Work in Mexico states:

> Choosing Juan Chavez as his traveling companion, Powell traveled horseback to Monclova, Musquiz, Progreso, and Villa Juarez to the north, Patos and Viezca to the west, Ventura, San Salvador, and El Salado in the states of San Luis Potosi and Zacatecas to the south; and to Galeana, Rayones, and San Rafael to the east. When there was no building available in which to hold services, Juan Chavez held a flaming torch aloft while Powell preached to those who gathered around.[82]

Madero Institute

While William was traveling on his evangelistic trips, his wife, Florence Mayberry Powell and her sister Annie Mayberry started and ran a school for girls in their home in Saltillo. At that time there was little education for females in Mexico. Their school became popular, and with William's government contacts that one school multiplied to schools in numerous locations.[83]

In 1883, after he had been in Mexico only one year, William travelled by train to Richmond, Virginia, in the United States. The Mayor of Saltillo and the Superintendent of Public

Instruction for the State of Coahuila accompanied William at the request of the Mexican governor. Governor Evaristo Madero wanted to offer to the Southern Baptists properties on behalf of the state for an orphanage and girls' school that William desired to start.

However, because of issues regarding separation of church and state, the Southern Baptists would not accept the gift outright. It was decided to set up a board of trustees to buy the building for $10,000. The Secretary of the Southern Baptist Mission board, H. A. Tupper, then travelled to Mexico to carry that out. The school was named after Governor Madero. Tupper's daughter became one of the excellent teachers there.

Starting in 1885,[84] when Madero Institute opened, Senator Jose Cardenas, a prominent Mexican state official, served for years as principal and teacher at the Madero Institute.[85]

For many years that school was very successful. Within two years after the opening of the Madero Institute, six female graduates were employed as native missionaries. Ironically it was the Madero Revolution of 1910 (incited by the grandson of Governor Madero) which forced the school to close in 1914.[86]

Cooperation with Other Missionaries

It was not until 1884, after the Powells and Annie Mayberry had been in Mexico two years, that other American missionaries began joining them. That same year William started the Coahuila Mexican Baptist Association consisting of eight churches with combined membership of 150.[87] Also in 1884 his Alma Mater (Union), which at that time was known as Southwestern Baptist University, awarded him an honorary Doctor of Divinity degree.[88]

From the beginning William cooperated with other evangelical mission groups. In 1845 the Southern Baptists had broken off – over issues of the Civil War, including ownership of slaves – from those who came to be known as the Northern Baptists.[89] However, by the 1880's that was behind them. Surely William, dedicated to his spiritual mother, the former slave Aunt Laura Ward, never believed in slavery. In the late 19th century the two Baptist groups were still almost identical in doctrine and in Mexico they worked in cooperation.

At the dedication of the Northern Baptist church building in Monterrey in 1885 Thomas Westrup offered prayer and William Powell, who served with the Southern Baptists, gave the dedicatory charge. The Northern Baptists described William's message as a "sermon of much power and truth."[90]

For that meeting 126 dignitaries travelled to Mexico from the United States, including several Northern Baptist representatives.[91] Perhaps the reason William was asked to keynote and not a Northern Baptist dignitary was that William had helped heavily in raising the funds to build that church and in securing the architect.[92]

Two years later at the dedication of the building of the Northern Baptist church in Mexico City, Thomas Westrup gave the sermon and William Powell offered the prayer. $18,000 was required to build that second edifice. Philanthropist John D. Rockefeller gave $7000 for the lot.[93]

Photo includes both Northern and Southern Baptist missionaries and had to have been taken prior to 1887 when due to poor health Flournoy resigned. Top row F. T. Trevino, W. M. Flournoy, Porfirio Rodriguez, Bottom row T. M. Westrup, W. D. Powell, O. C. Pope, W. H. Dodson, J. D. Wright.

Dangerous Trips

William took long, sometimes dangerous, trips on horseback through the Mexican countryside and mountains. Florence Powell Harris wrote:

> Father would start out on his preaching tours with his saddle bags filled with Gospel portions and tracts. He humorously said that sometimes his… [horse] would stumble and the literature would fly in every direction, but this too, was a means of spreading the Gospel.[94]

A Century of Baptist Work in Mexico stated:

> The inter-city highway system [in Mexico] was not launched until 1925. Travel was mostly by horseback or by wagon on dirt roads and trails.... from 1876 to 1911 nearly 15,000 miles of rail lines were built, connecting the principal cities of Mexico.[95]

William wrote:

> While preaching in the new church at Saltillo one night I saw three young men who had evidently heard for the first time the story of Christ's dying love. I hurried down and greeted them and they told me that in their country the priests forbade their reading the Bible, saying that it was the meanest book on earth. They ... had never heard such good things as God's love for sinful men [told that night.]...and that the song "Jesus Saves" was so sweet that they did wish their mother and father could hear it.
>
> It was arranged that I should go in a few weeks to their home. In doing so I had to swim my bronco through a swollen stream and I reached there in wet clothes....I was taken into a room....
>
> The father and mother and Manuel, who had accompanied me, assembled in the patio ... and engaged in an earnest conversation.
>
> Sixty people were murdered while the Gospel was being established in Mexico and some were killed not very far from this ranch. From the ... manner of their conversation, I felt assured that the people had threatened that if I preached they would kill me and kill the family....I never prayed more earnestly for divine help.
>
> Finally the old mother came into the room with some clothes on her arm saying: "Senor, you must remove the wet clothes or you will die of pneumonia. The people are already beginning to assemble...I believe that everyone in this community is coming to hear you preach"....she closed the door and departed.
>
> Mexicans are small of stature and they could find no clothing large enough. I had soon...put on the under garments and began the task of adjusting the outer ones.
>
> The pants were yellow and made to fit skin tight. With the greatest difficulty, I forced them on and with still greater difficulty was able to fasten them.
>
> Then I put on the green and white striped vest, and when I had buttoned it up I found that there was about twenty minutes' recess [a serious gap] between the bottom of the vest and the top of the pantaloons.

> ....I put on the white blouse, the sleeves of which came just below my elbow and the tail of it came nearly to the waist band of my pantaloons....
>
> The shoes were yellow gaiters with heels fully six inches high and with points after the pattern of a tooth pick. They were entirely too small, but... I succeeded in getting them on and to my horror discovered about eight inches of white uncovered space between my pantaloons and shoes. [Country folks in that region did not use hose]. I am sure that I looked more like a clown than a preacher.
>
> I hurried across to where the people were assembled. Mexicans, as a race, are very polite, but my ridiculous garb was more than they could stand. They broke out into laughter and the longer I stood there the worse the situation grew. The old folks rushed out and brought in the dining room table, placing it in front of me and Manuel covered it over with a bed blanket, which he draped over the side. They wished to hide all of me that was possible.
>
> One would have thought this ridiculous garb would have destroyed all seriousness for the evening. The room was filled to suffocation. I read the Scripture, prayed and sang the song "Jesus Saves." I took as my text John 3:16....["For God so loved the world that He gave His only begotten Son, that whoever believes in Him should not perish but have everlasting life."][96]
>
> In a few moments I had forgotten all about the clothing as did the people also. Conviction of sin came to their hearts. While I was yet preaching, the old father stood up and said... "I see it, I see it, I see it" and then and there publicly confessed Christ. Before we closed the services, twenty-three people had confessed a hope in Christ. Two of those boys are today leading Baptist ministers and have been eminently successful as soul winners.[97]

One of William's yearly mission reports stated:

> I was violently opposed by the ignorant and fanatical ranchmen. My life was threatened daily. Once, while at prayer, a ruffian attempted to plunge his dirk [dagger] through me, but was prevented by one of our deacons, who... usually accompanied me. I have been shot at, the ball passing harmlessly under my horse. I have been searched by a highwayman, who after making an inventory of the assets of a traveling Baptist preacher, offered to lend me enough money to get home on. I thanked him, but declined the offer.
>
> A fanatical officer, without any cause, detained me a prisoner for two days, and then sent me one hundred miles under guard, thinking vainly that in this way he would intimidate me and hinder the work.
>
> I left a Bible at a ranch where I had been preaching. Many of the people believed and some were baptized. The overseer and owner became furious.

> They put up the Bible as a target, and shot at it one hundred times. Only two balls entered the sacred book. The people were then all convinced, saying "Surely that is God's book, or Don Manuel would have put every ball through it."
>
> One who had previously been a fanatic begged that the mutilated Bible be given to him, and the reading of it led to his conversion. Truly the Lord can make the wrath of man to praise Him.[98]

In another account William stated:

> A man came to the house where I was stopping at Cienega, pistol in hand, to kill me but I took a Catholic New Testament and showed him…and he was ever after a true friend.[99]

The report of the Mexico Mission for the Powell's third year states:

> Since last May the Saltillo church has received eighty-seven members. Seven of the members are preparing for the ministry, one…is at Baylor University, two are at the Seminary in Louisville, and four are in Brother Powell's Theological class.
>
> ….A good brother in Georgia sustains three …, who are doing a grand work as "forerunners," distributing hundreds of Bibles, and portions of the Scriptures, and thousands of pages of religious tracts. The Dunn department of the Madero Institute has nineteen orphans.
>
> A revival in the college chapel resulted in thirty-two accessions [new members] to the church.[100]

The mission report in 1886 was not written by William but discusses his work on the ranches:

> Several years ago Edward Lara, aged seventy-eight years, was instructed in the Gospel by Brother Powell, who was the teacher of Lara's son. Some… years before he had received a Bible [from American soldiers during the Mexican-American war in 1846[101] – 35 years prior], in whose contents he became much interested.
>
> Being a man of large possessions, he invited our missionary [William] to preach on his ranches; the result of which was the baptism of Lara and seventy-four other…converts and the organization of two churches.
>
> Mr. Bustamante, an ex-governor…requested our missionary to visit his ranches, of which Brother Powell writes: "These ranches and haciendas cover a territory nearly equal to the State of Tennessee, and give us an opening never equaled in this country…" The authority given to Brother Powell is worthy of record.[102]

That report goes on:

> The work of the church and the school is enough to fill the heart and the hands of our brother [William], but he has, in addition, the care of all the churches of the State and the responsibility of vast tracts of country calling unceasingly upon him to come over and help them with the Gospel...He must and will be reinforced.[103]

Inspiring Mexicans to Serve God

A Century of Baptist Work in Mexico stated:

> Powell was adept at persuading nationals to dedicate themselves to mission work. In 1887 Alejandro Trevino Osuna was 19, an engineering student.... Powell persuaded him to leave his studies and go to San Rafael to establish a center of evangelization.
>
> When Trevino arrived at Saltillo, Powell was on one of his trips and had left no word for Trevino. Trevino, however, continued on alone to the mountain village of San Rafael and presented himself to the principal rancher, saying that Mister Powell had sent him to preach the gospel.
>
> The old man replied: "We don't know Maestro Paulo and we don't know what the gospel is, but come in."
>
> Learning that there was no school in the community, Trevino offered to establish a school. By that means he was able to support himself and had opportunity to preach.
>
> On July 17, 1887 Trevino baptized 59 candidates and organized a church at San Rafael. On the same day he organized a church with twenty members at San Joaquin. Trevino, during a ministry of more than fifty years became Mexican Baptists' most beloved and honored citizen.[104]

From his first year in Mexico William wrote of a desire to start a school to train young men studying for the ministry.[105] That dream was realized when Center for Theological Studies, later called Zaragoza Institute, was established in 1889.[106]

The Mexico report for 1889, after William and Florence had been in Mexico seven years, stated:

> I have been driven away from ranches because I was a preacher of the gospel. I have traveled over lonely mountain trails, where for a fortnight I would see no road for wheeled conveyances. On one of these routes twenty-three men have been shot since I began these periodical visits. I have slept on

> prairies, on the mountainsides, in hovels, and in the homes of the wealthy. God has wonderfully blessed my poor efforts in this direction.
>
> The other day I announced preaching under a natural bridge, along one of these mountain trails, and seventy people assembled and listened with tearful earnestness to the story of Christ's redeeming love. One beautiful Sabbath day in July, 1887, in a ranch called San Rafael, I baptized fifty-seven people. It was the happiest day of my life.
>
> We have carried the work from the Texas border to the Pacific coast. Opposition is waning. Now, I almost universally meet a warm welcome. The government gives us full protection....
>
> We have eighteen American missionaries and fifteen native workers. Our force of workers is insufficient to occupy the territory already open to us. There are eighteen organized churches and some 600 members. "Truly this is the Lord's doing, and marvelous in our eyes." [Psalm118:23][107]

Years later a report came regarding that natural bridge meeting when a man who had been a noted bandit stated:

> "Twenty years or more ago I was in Rayones and there was a Baptist minister there named Powell. He was to leave the next morning for Galeana. He had baptized a number of people through all that region and was organizing churches and I determined to follow him and kill him. When I overtook him he and his companion Charez [sic- Chavez] began to talk to me about my soul, gave me a New Testament and made me promise to come and hear him preach that night. I did so and was deeply stirred about my soul.
>
> "I came home, read the New Testament and have read it in my family and prayed regularly to God for years and have realized the forgiveness of my sins; but I had lived such an outrageous life that I have never had the courage, until now, to come and tell you what the Lord has done for my soul, and to ask you to receive me and my family on our experience of grace as candidates for baptism."
>
> [Powell stated:] "When this was told to me I remembered well the occasion and the earnest talk we had on the natural bridge where he had intended to murder us. If I had not witnessed for Christ I would certainly have been killed. Twenty-eight people were murdered along that route that year."[108]

Florence Powell Harris, my grandmother, wrote that about 1890, when she was about a year old, her father (William) took a trip to the Holy Land and stopped in London on the way. Famous preacher, Charles Spurgeon, had William preach twice that Sunday at his church. His preaching at Spurgeon's church is also shown in the biographical data for William Powell maintained by the Southern Baptists.[109]

William tried to get his wife to accompany him on that trip to the Holy Land, and she wanted to go. However, little Florence was only one year old, and mother Florence did not want to leave her baby. Shortly after William left, a growth was discovered in the baby's eye. On the doctor's advice mother and child immediately went by train to Austin, Texas where emergency surgery saved the infant's sight.

Health Problems of Early Missionaries to Mexico

Unfortunately, both Mr. and Mrs. William Flournoy, after serving only six years as the first official Southern Baptist missionaries to Mexico, had to resign in 1887 for health reasons. They never regained their stamina. She died in 1890 and he in 1892.[110] Neither reached age 45.

Over a span of time various Baptist missionaries had to permanently leave Mexico due to decline in their health or that of family members. Among them were the Samuel Gorman family, P. H. Goldsmith, Sara Hale, Robert Whittaker, and the A. C. Watkins family.[111] Numerous others had to take temporary respites to regain wellness.

A Century of Baptist Work in Mexico states that in 1892 William was called back to the US for fundraising.[112] My grandmother, Florence Powell Harris, indicated on that occasion her parents left her in Mexico in the care of her mother's sister and co-laborer, Annie Mayberry. Earlier Annie had returned to the US to recover from an illness. Florence Harris stated:

> Aunt Annie kept her own house, so mother permitted her to keep one or the other of the little Powell girls as a companion, most of the time. When time came for the Powells to return to the United States... Aunt Annie entreated Mother to allow me to be left with her.[113]

At first Florence's mother refused to leave baby Florence with her sister but finally gave in. Florence Harris continued:

> Scarcely had my parents reached Murfreesboro, when a cable arrived saying, "Miss Mayberry is desperately ill," and this was followed shortly with another message, "Miss Annie has erysipelas" [a bacterial skin infection]. The third telegram...read "Miss Annie is dead." My mother...had double distress, for she not only had lost her sister but also had no inkling as to the whereabouts of her youngest child! Father rushed back to Mexico to see after my aunt's affairs and to bring me back to mother.
>
> The only recollection I have of that brief and tragic stay is of my crying and pounding on the glass partition that separated me from my aunt who was desperately ill on the other side. Surely God used a tiny little girl to comfort the heart of a lonely young woman for a short while. Aunt Annie is buried in Toluca, Mexico.[114]

Annie Mayberry died October 9, 1892 in Toluca. She was 31 years old. Little Florence would have been three years and two months old then.

Years earlier Annie had struggled in learning Spanish but by the time of her death had been in Mexico 10 years. James Chastain in *30 Years in Mexico* stated about Annie, that she: "spoke Spanish like a native, greatly useful as teacher and otherwise....We cannot understand it [her death]."[115]

Annie's obituary in the December 1892 *Foreign Mission Journal* stated:

> Miss Maberry evinced the patience, fortitude and self-denying love, which subsequently, were so characteristic of her life..... Love and admiration for her was universal among missionaries, teachers, children, neighbors and citizens. No jar of discord ever disturbed her harmonious relations with any with whom she was associated. The common people almost idolized her. Their pet name for her was "Anita, la simpatico" – "little Annie, the lovely."[116]

Annie Mayberry.

Another promising young missionary, Marion Gassoway, died December 12, 1895 of typhus. He was 30 years old and had been in Mexico only two years.[117] Alone in his station at the time of his death, he was attended during his final hours by an American Presbyterian missionary who wired Powell regarding Gassoway's passing. On receiving word, William immediately went to handle burial details.

William wrote that the Presbyterian missionary indicated that when Gassoway was diagnosed with typhus he stated: "I am in His [God's] hands."[118] Earlier that year William had high compliments for Gassoway when together they went on an evangelistic trip. During that trip William had been quite ill.

William stated in a letter to the foreign missions director in Richmond:

> In April I was with brother Gassoway when it seemed that I was not long for this world and Brother Gassoway was so strong and healthy. Now my health seems to be perfect. My brother, how little we know when the Master shall call us. Let us live ready to go any moment.[119]

Because of the distance between mission stations and the difficulty in travel, the various missionaries were not able to assemble for the burial. In *Thirty Years in Mexico,* fellow missionary James Chastain wrote about visiting Gassoway's grave in Zacatecas a few months later. Chastain indicated on that occasion he was accompanied by William Powell and one other missionary. Chastain wrote:

> With tear-bedimmed eyes we thank God for the zeal, piety and sweet spirit of our brother [Gassoway] and plead that others may come and take up the burden he laid down.[120]

The Powell family was also repeatedly affected by illness. *History of the Baptists of Texas,* tells that in June 1888 W. D. Powell went to San Antonio, Texas for medical treatment of one of his children. While there: "Dr. W. D. Powell labored for some weeks among the Mexicans of that city."[121]

Books on Baptist history in Texas still mention the revival that happened in San Antonio at that time. Leon McBeth in *Baptist Heritage* reported that in 1888 a revival broke out among the Mexicans in San Antonio, Texas as a result of W. D. Powell's preaching.

In one annual mission report William wrote that for the first time in two years he felt well. His wife Florence was also frequently ill. From reading William's private correspondence to Richmond it appears that during the Powell family's last four or five years in Mexico he was ill more often than he was well. However, he kept pushing his body to the limit to go on long evangelistic trips. Twice he had to leave Mexico at doctor's orders to recuperate. On the second such exit he was so ill that five doctors were called in to try to help him.[122]

The Powells Move to Toluca

In 1892 after returning from their time in the United States the Powell family was sent to live in Toluca, only about forty miles from Mexico City.[123]

The 1892 mission report presented "Brother Powell's theory of missions" as accepted policy and stated:

> As a general evangelist, Brother Powell visits all the missions and preaches at stations not belonging to any of the missions.

> Brother Powell is now at Toluca, the capital of the State of Mexico, where he edits the Baptist paper LA LUZ, and is more centrally located than formerly for his broad and growing work.[124]

That report stated that Powell wanted to expand even further and go into Mexico City, where at that time there was no evangelical work. The document also discussed a recent meeting of missionaries from various Mexican outposts: "The utmost harmony prevailed in the meetings, with a striking manifestation of the divine presence."[125]

As he had done in Saltillo, William soon raised funds for and started an orphanage in Toluca.[126]

Effective Ministry

Several of the Mexican mission reports describe people being moved to tears during William's meetings. The power of God on him was evident – surely the result of the prayers of Laura Ward, the former slave.

William had no hesitancy in loving the Mexican people. He was interested in their individual lives and in their history. He was curious about the origin of the Aztec civilization and enjoyed talking to an American friend in Mexico who studied that history. Although he gave little time for that matter, his interest led to the conversions of some of the nationals connected to that research.[127]

William believed in the capabilities of the Mexicans. His goal was for their churches to be self-sufficient – free from foreign support. In 1894 the church in Saltillo was the first self-sustaining Southern Baptist Church in Mexico.[128]

Mississippi Baptist Preachers, by L.S. Foster, published in 1895 contains an inspiring four page article about William. It concludes:

> His energy is tireless, his zeal unquenchable, and it is impossible to estimate the amount of good he has done in Mexico in letting in upon this people the blessed light of the gospel of Jesus Christ. He may be very appropriately styled the apostle of Mexico, for he has the true apostolic spirit.[129]

The Powell's notable guests in Mexico included famous evangelist and founder of Moody Bible Institute, Dwight L. (D. L.) Moody, and his Gospel singer, Ira Sankey.[130]

Little Florence, also recalled as a young child sitting on the lap of famed orator William Jennings Bryan when he visited her family in Mexico.[131] Bryan was first a US Congressman, then three-time presidential candidate, and later Secretary of State under President Woodrow Wilson.[132] He visited Mexico, including Toluca, in December 1897.[133]

William Powell is on back row far right. William Jennings Bryan is in second row second from left.

The Conference on the Holy Spirit

The pinnacle of William's ministry in Mexico was most likely the conference on the Holy Spirit conducted by D. L. Moody and his song leader Ira Sankey in 1895. William, having experienced the blessing of the Holy Spirit in his ministry, wanted that power for all the evangelical missionaries of various boards in Mexico. Therefore, through his connections he arranged for Moody to travel there.

According to Chastain, who attended that conference, William personally financed that event and entertained the guests.[134] In William's private correspondence to Richmond he indicated that he had from his own funds secured three hotels for the missionaries and other guests. All was free to the attendees including streetcar fare.[135]

Chastain described that series of meetings:

> In 1895 conditions were ripe for a great soul-winning campaign that would sweep all over the country. Accordingly it was decided to hold in Toluca [where the Powells then lived] a series of inter denominational conferences. Moody and Sankey still in their prime accepted an invitation to be with us.
>
> The roster showed 132 messengers coming from every part of Mexico and representing perhaps a dozen boards. These with the visitors and local attendance gave us immense congregations. Besides Moody's daily sermons, two dozen strong papers were presented, some in English other in Spanish, touching the presence, personality, power and work of the Holy Spirit.
>
> ....The last meeting was a *veritable Pentecost*. The Mexicans present who could not understand English were amazed and dazed. The people wept, laughed, shouted and embraced each other, beside themselves with joy. The influence was permanent and far-reaching and marked an epoch in Mexican missions. Through succeeding years thousands were converted to God.[136]

At that time William reported to his supervisor in Richmond that there were six to eight hundred people present in the large theatre housing the Spanish language portion of Moody's conference.[137]

Trials and Tribulations

During the last few years of the Powell's ministry in Mexico a series of hardships came on his family akin to the trials of Job in the Bible.

Part of the problem was that the Southern Baptists, who were relatively new in foreign mission work, had not appointed a field director for Mexico. As the senior missionary William was recognized by the nationals and by other denominations as the leader of the group. However, the foreign missions headquarters in Richmond had not officially given him that title. Some of his fellow missionaries resented it when William received such recognition.

Furthermore, William was out performing every other missionary. He did not do that to upstage anyone. He labored simply from devotion to God and love of the Mexican people. He was a charismatic man and many people both in America and Mexico loved him in return.

Often in a work environment when one person outshines everyone else, jealousy ensues. That happened at that time among some of the missionaries.

William was grieved when he learned that even missionaries in Japan were jealous of the attention the Mexican mission garnered. In reply William tried to explain that he and his family were only trying to serve God and had not been extravagant in spending. They had always lived simply in native style houses[138] and he was constantly looking for ways to save money for the mission.

Annual Missions Meeting, Saltillo, Sept. 1895. In back row left to right are Powell, Watkins, McCormick, Gassoway (turned sideways), Dr. Willingham from Richmond. Gassoway died three months later. Steelman is in front row far right.

Mr. Steelman, a new Southern Baptist missionary, upon reaching Mexico refused to go to the location he had been assigned by the main office in Richmond. Without consulting anyone, Steelman took another town of his own choosing. With no supervisor on the field some others were violating multiple mission policies, taking extended trips back to the States, and doing little work. When William reported those infractions to Richmond, he became a target.

After he realized that he had been reported, Steelman worked to incite other missionaries against William. It caused such a ruckus that the Foreign Missions Director traveled by train from Richmond to Mexico. Ultimately Steelman was fired, but by then he had already incited hard feelings in other missionaries against William.

Some other missionaries then revolted against Steelman's firing. William, not one to bear a grudge, went the extra mile by helping Steelman and his family to get their belongings to the steamer for their voyage home at the time of their exit.

Shortly thereafter William sought and received permission to make a quick trip to Texas after he learned that the prominent judge with whom he had entrusted his assets in Texas had embezzled others. William was most fearful that his own name would be tarnished should his power of attorney be invoked in a deceitful manner by that judge.

A good portion of the Powell's private assets were lost, but he successfully revoked his power of attorney with the judge. During that short trip William was also able to visit his father and half-brother who were living in Texas. That was the last time he saw his father, who died the next year.

Meanwhile another Southern Baptist missionary to Mexico was jailed. Chastain explained in *Thirty Years in Mexico*:

> [Missionary] H. R. Mosely [sic–Moseley] wrote in English a tract on "Three centuries of Romanism [Catholicism] in Mexico," which was mistranslated and garbled by bitter enemies causing his imprisonment and endangering his life. Following the dictates of prudence he withdrew from Mexico.[140]

Upon William's return to Mexico from confronting the judge in Texas, both the governor of his state and the President of Mexico contacted William directly to find out what had happened concerning Moseley.

Instead of being peace loving, Moseley had apparently made enemies in Mexico who purposely mistranslated Moseley's text. Both the Mexican government and the Catholic Church felt insulted. The newspapers published the story inciting the Mexican populace to anger.

William smoothed things over by instructing Moseley to issue a written apology to the Mexican government. Mexican officials wanted Moseley to serve out his sentence in a less than desirable Mexican jail, but at William's request the President of Mexico ordered Moseley's immediate release. Mosely was advised for reason of personal safety to immediately and permanently leave Mexico.[141]

Around that time the Southern Baptist Mission was low on funds. Because of their debt, the Board did not pay all that had been promised to Powell. This worked a hardship on the Powell family who at that time had three children in college.

William offered to resign and take a job but was not allowed to do so. Meanwhile, he had on faith promised to personally cover the expenses of the Moody conference for which he had rented three hotels. Somehow, he was miraculously able to meet those expenses.

William continued on his long missionary trips. On one trip in April 1895, almost immediately after the Moody conference, he wrote that he was far from well, but was 700 miles from home. He would stay in bed during the day then get up to preach at night. Word from his wife had reached him that at home their baby Henry Taylor was seriously ill.[142]

Three days later in a letter William stated he was feeling somewhat better. He was thankful that he had sold 1500 gospels in the last two weeks.[143] However, within a month, again William was "far from well."[144]

William's health was in a precarious state. After the Conference on the Holy Spirit, Moody and Sankey wrote to William four times[145] begging William and Florence to go to Moody's medical retreat at Northfield Sanatorium in the States for a month's rest and recuperation at no charge.

William repeatedly requested permission from Richmond to go to Northfield. It was never granted even though William wrote that his doctors had been advising him for more than a year that he take such a respite.[146]

Although he was not healthy, William continued preaching. In a letter dated July 2, 1895 he stated that his health was "so, so," and that he was "head over heels in work." He had preached in the state of Hidalgo, "the first Baptist sermon ever in that state." In the morning in Hidalgo he preached to Mexicans and in the afternoon to 150 Englishmen.

He did not seem to bear malice that Dr. Willingham, his supervisor in Richmond, had refused him permission to go to Northfield. At that time William's half-brother, a sheriff and tax collector in Archer County, Texas, was in Mexico to visit.[147]

Joseph Daniel Powell half–brother of William. Joseph was a sheriff in Texas and had 12 children.[139]

After spending eleven days ministering among the ranches with Florence, William stated in a letter dated November 18, 1895: "I am longing for more of God in my thoughts and His Spirit in my heart."[148]

Upon returning home they learned of an enraged woman who wrongfully believed that William was responsible for her dismissal.

While none of the missionaries had taken a vow of poverty, some felt it wrong for William to own property – so he sold his home. William wrote that his wife had reluctantly consented but was not pleased. William added: "Pray that the power of the Holy Spirt be upon me. He knows I love Him and would not neglect Him."[149]

At that time William told the Richmond office he still owned a piece of land he held for his children and stated:

> Nothing shall stand in my way of loyal, faithful service to Christ. I believe they [those who thought it wrong for him to own real estate] were wrong but have given them the benefit of doubt….I crave to be useful as a winner of souls.[150]

Eleven days later William received word that his beloved fellow missionary, Gassoway, had died. Upon learning that news, Dr. Willingham in Richmond asked William for a recommendation for a candidate to replace Gassoway at that station. A few days later William sent Willingham the name of Robert Mahon, who was appointed to Mexico.

Throughout 1896 William experienced much illness. At one point he involuntarily lost 18 pounds in two months. At times he went "from bed to the pulpit." His requests to leave Mexico for a brief respite fell on deaf ears. His reply was: "I will remain with the work unless I break down entirely."[151]

When repeatedly he was not paid full amounts promised, he seemed more concerned that the Southern Baptists were in debt than that he had been slighted. "God will provide," William responded.[152]

That year William was troubled by some members whom he said were "seduced by spiritualism" (speaking to the dead), a practice which was clearly against Baptist doctrine. That challenge was mixed with experiences of joy. In the same letter he seemed excited that his children, who had been away at school, would be coming home in June.[153]

When his son Willie and daughter Annie came home from college they both became active contributors to the ministry. William wrote:

> Willie was elected superintendent of our Sunday School. Annie is organist and teaches a class. It fills my heart with gratitude to see my children so deeply concerned for the welfare of Christ's cause. They visit the sick and find many ways in which to lend loving help.[154]

Finally at the end of 1896 William was granted a leave of two months. He previously wrote "I must go where temptations to speak in public will be few and far between."[155] However, on reaching the States almost immediately he spoke five times in one Sunday and continued speaking. He wrote in the same letter: "I am distressed because I do not improve more rapidly."[156]

On December 17, 1896 William wrote that for two months he "feasted on the fat of the land in Tennessee, Alabama, Louisiana, and Kentucky and spoke in 20 places." However, in Jackson, Tennessee he was stricken with what the doctors diagnosed as typhoid or protracted malarial fever. That was followed by flu which left him very weak and threatened by jaundice.

By the end of the year and still weak, he was back in Toluca, Mexico where he wrote: "I am praying for close fellowship with the Holy Spirit during the next year (1897)."[157]

William did have close fellowship with the Lord during 1897 but for reasons he never expected. In January 1897 he was approached by three fellow missionaries together who were jealous over the fact that they had seen William's name in print in US newspapers during his recent visit there.

They were especially enraged that one US newspaper reported William as "the head of the work in Mexico." Of course, that was untrue since even though the Southern Baptists had repeatedly used William's services in that capacity, they had never officially appointed a "head of the work in Mexico." William tried to explain to the three missionaries that he had made no such claims himself. Furthermore, he had not seen the article until a week after it came out – too late to ask for a retraction.

Thus ensued a series of angry accusations including the missionaries saying that William had lost the confidence of the Mexican people. They remained angry that William had notified headquarters about Steelman and others who refused to follow policy. The missionaries said that if William did not resign, they would all three themselves resign.

William was shocked by what they said but did not agree with their analysis of himself. That same day he stated in a resignation letter to Richmond that since the others were all "good men" it would be better for the mission to lose one man (William) rather than three. Even in that letter William proclaimed his desire to honor God.

That same day Thomas Crittendon of the U.S. consulate in Mexico heard of William's planned resignation and sent a four page hand written letter (now in the archives of the Southern Baptists) begging William not to resign but to stay at his post. Crittendon cited William's "willingness to do whatever is prompted by wants of humanity." That missive stated: "....we know the right man in the right future when we see him."[158]

Newspapers in Mexico were quick to pick up the story. One article in English headlined "Mission Row – Jealousy Believed to be the General Cause." Another article also referred to it as "A Mission Row." The concluding paragraph of that stated:

> Dr. Powell has...always had the reputation of being an earnest worker,...a kind and genial gentleman and perhaps a trifle too broad minded and tolerant to suit the tastes of his fellow workers.[159]

MEXICAN HERALD THUR

A MISSION ROW.

BAPTIST MINISTERS RESIGN IN A BODY.

The Cause of Their Action Is Their Discontent With a Member of Their Own Church.

There is a serious row in the ranks of the missionaries of the Southern Baptist Conference, who are established in Mexico. The result is the resignation of three out of the five missionaries and the probable resignation of a fourth. Those who have sent in their resignations and are now packing their household goods with the intention of leaving shortly for the United States are Rev. [illegible] of Saltillo, who is in charge of the large girl's school in that city in addition to his other missionary work, Rev. H. P. McCormick, of Morelia, and Rev. A. C. Watkins of Torreon. The other missionary who has not yet resigned but who is known to be in sympathy with the other three is the Rev. J. G. Chastain, of Doctor Arroyo, N. L.

The trouble which has caused the resignation of the missionaries had its beginning a long time ago and has been growing every day until the outbreak which occurred two months ago. It is over the Rev. W. D. Powell, who has charge of the missionary work at Toluca and who is know as one of the most prominent Baptists in the republic. Dr. Powel is charged by his brother missionaries with a number of sundry misdemeanors, the most prominent of which are untruthfulness, and exaggeration, which, in their eyes, make him unworth of representing his church in the missionary field. There are number of other charges which are lodged, the exact nature of which cannot be determined, because the Board of Southern Conference of Baptist under whose direction the Mexican missions are conducted refused to try Dr. Powell on the charges which were presented to them, the other missionaries have sent in their resignations.

Several months ago the missionaries who resigned held a meeting and decided that they would present charges against Powell before the board of missions of their church. They began corresponding with the secretary of the board and presented their charges in writing.

They made particular mention of the fact that Dr. Powell was in the habit of greatly exaggerating his work, that while he had charge of one of the smallest mi[illegible] in the Republic, the published rep[illegible] sent in by him showed that the work accomplished was greater than that of any other mission. He was also charged with mixing up religion with business and using opportunities that he secured through his position as a clergyman to sell stock in land companies in which he was interested. After considerable correspondence the Board decided to allow the accusers to bring their witnesses to the United States where they would be examined and the whole matter sifted to the bottom.

After considerable delay sufficient evidence was secured and the witnesses dispatched to Richmond, Virginia, where the Board was to hold its session. After they had reached Richmond, much to their surprise they learned that the Board had reconsidered the matter and decided not to let the charges come to a trial. So the witnesses returned to Mexico thoroughly disgusted.

In the meantime Dr. Powell, who was thoroughly posted on what was going on decided to make the trip to Rich[illegible]

[illegible] The board [illegible] only right that [illegible] clear himself and commenced an examination into the charges, without however having the witnesses present. The result was that the charges were dismissed and the Rev. Doctor entirely exonerated.

When the action of the Board reached Mexico and it was found that Dr. Powell had been exonerated, without his accusers being given a chance to prove their charges, the latter were anything but pleased in fact they were very indignant that their word should have been so lightly regarded. The outcome was that they determined to sever their connection with the church as it would be impossible for them to work in harmony with Dr. Powell.

When the matter is finally investigated, as it shortly must be, it will most likely be found out that the trouble came from the fact that there is no head of the Baptist work in Mexico. There was a sharp rivalry between the missionaries on this account which has undoubtedly led up to the trouble. When the work was first organized in Mexico in 1883, Dr. Powell was practically the head of the work and was regarded as its superintendent. This state of affairs did not last long as it is one of the principles of the Baptist chuch that one minister shall not be higher in authority than another. Since that time, however, Dr. Powell has kept a sort of control over the work in the whole republic.

The Baptist church in Mexico is represented by the Northern Baptists and the Southern Baptists, two entirely distinct bodies. The Northern Baptists, who are represented in Mexico City by the Rev. Dr. Sloan, are not mixed in any way with the row, as they are an entirely separate organization.

Dr. Powell has been in Mexico since 1883. He has always had the reputation of being an earnest worker, a kind and genial gentleman and perhaps a trifle too broad minded and tolerant to suit the tastes of his fellow workers.

That was soon followed by numerous letters to Richmond from various Mexican public officials and businessmen highly praising William and his upright business dealings.[160] Some of these individuals wanted to collect many signatures on their letters to Richmond, but William would not allow it.

The highly regarded Mexican evangelist Alejandro Trevino wrote to Richmond of William's love for the Mexicans and stated:

> Among Mexican evangelical worker he [William] is one of the most beloved. We would all like to see him on our fields because when he visits us a spiritual revival is always seen in our congregations. He does not know I write to you.[161]

At that point William's wife Florence, far along in a pregnancy, was considerably worried about the whole affair against her husband.[162] William wrote:

> The dear Saviour be very near us all and may His mind and Spirit fill us. He knows that I feel so unworthy that I do not wonder that others criticize me.[163]

A few days later (February 26, 1897) William announced that the previous night God blessed their family with the arrival of twins – a girl and a boy (Ethel and Ernest) – their ninth and tenth living offspring. William then wrote: "The cause of my Master is dearer than all else. I can gladly suffer all things rather than see His cause suffer."[164]

The twins, born almost 21 years after William and Florence's first child, joined the other Powell children – William Eaton (Brother) Powell (born in 1876); Anita (Annie) Powell (1878); Nell Powell (1880); Charles Breedlove Powell (1884); Mary Mamie Tupper Powell (1886); Paul Powell (1887); Florence Powell (1889); and Henry Taylor Powell (1894).[165]

The Southern Baptists did not immediately accept William's resignation, but rather chose to conduct an investigation into the matter. Despite being asked three times to put any complaints into writing, the three accusing missionaries declined to do so.[166]

However, William's life was still put under a microscope. In relating how he owned property in Mexico William explained he had some money from the sale of his former residences in Texas that he had used to buy property in Mexico. Fortuitously he had moved part of his money to another bank in Texas before the judge absconded with the rest.[167]

On May 15, 1897 William wrote: "The Lord was never more precious to me and never did I have more longing to spend and be spent for His service."[168]

In June Willingham and another man travelled from Virginia to Mexico to investigate the charges against William. William was 700 miles from home and more than 100 miles from a railroad on an evangelistic trip when he learned that Willingham was on the way, but he wrote that he would leave at once for home.

Within a few days of his arrival to Mexico Willingham seemingly smoothed over issues with the complaining missionaries and convinced William to withdraw his resignation.

Then another crisis struck. The Powell's second daughter Nellie had been away at Cox College but came home to visit. Not wanting to trouble her parents, she had not let them know that she had fallen down the stairs at college injuring her spine and kidney. She had run up medical charges but had refused to have the bills sent to them.

On October 4, 1897 the Powells heard some of the worst news parents could receive. The doctor told William and Florence that because of the extent of Nellie's injuries she could not possibly live.

William lamented that the family had lost nine years of Nellie's companionship while she had been in schools in the US. Family photos were taken then.[169] At that time the Powells' aching hearts could not have known that Nellie would live for many years to come.

Bottom row: Little Florence, twin Ernest, Henry, twin Ethel, Charles. Middle row Paul, Mother Florence, W. D., Mary (Mamie); Top row Nellie, William (Brother), Anita (Annie).

Despite the fact that Willingham had travelled all the way to Mexico to settle the complaints against William, some of the missionaries remained disgruntled.

When a Mexican newspaper published an uncomplimentary article about one of the three accusers, that man falsely believed William instigated it. Soon that missionary resigned.

William was on a committee but could not agree with other missionaries who wanted to abandon certain fields or turn them over to other denominations. Powell offered to pay the expenses himself to keep them open. Some missionaries wanted to discharge the Mexican workers. William's reply was, "We do not need any more American missionaries. We must look to the Lord to raise up men here."[170]

In January 1898 Powell wrote to Willingham saying he had used his own means to build a mission house that he wanted to give to the board, but he did not want anything said publicly about his donation.[171]

That month the second of William's accusers turned in his resignation. Even the man's best friend stated that lately the missionary had acted irrationally. William stated in a letter to Willingham: "I pray that God's will be done in this republic."[172]

Then, after hearing twisted reports, two other female missionaries sent letters to William telling him that he needed to resign. The second letter arrived while he was traveling. Florence, who often handled William's correspondence, retorted stating the Powells had no ill feelings against anyone – however, since all the trouble came from that one town, perhaps the other woman missionary herself should be the one to resign.[173]

Willingham continued asking William for advice concerning the Southern Baptist work in Mexico. William tried to explain Mexican laws and why the board should not deed Southern Baptist property to any missionary living in Mexico – because on death it would pass to his family rather than to the mission.[174]

In the 1898 yearly report William told of a group that had completely paid for their own lot and chapel. He stated:

> We only organize churches where the members agree to furnish a place for the meetings and current expenses. A member of the Toxco [sic Taxco] church recently walked one hundred miles to Toluca to get a supply of tracts for distribution and to tell me about the church work. My native helper [supported by a group in Texas]...recently walked more than one hundred miles on a missionary tour because he did not have the means to hire a horse.
>
> Some of our... native workers are the very salt of the earth. For piety and devotion they are unexcelled. The Lord raise up many more like them, that the good work may go on, and Mexico be taken for Christ and His truth.[175]

Meanwhile, William continued his evangelistic tours to distant locations. On one of his trips William took his daughter, Annie. One day they rode horseback for 45 miles partly through "*tierra caliente* where it was hot as an oven." William wrote that they traversed part of three states and Annie was "much fatigued."[176]

Later that same month (March 1898) William made a quick train trip to Richmond (which had to have been at the request of Willingham). There he stayed with Willingham and his family. Other missionaries quizzed him about that.[177]

William's letters praised the dedication of the Mexican workers: "Our members often manifest much of the apostolic spirit in their devotion to the cause of Christ."[178]

Then some of the other missionaries conspired to discharge or cut the salaries of Mexican workers – particularly those who had spoken up in William's defense. William, remembering one Mexican who had his salary cut and therefore could no longer help in the ministry, stated: "Poverty will not allow him to do more."

Meanwhile, the second missionary accuser, who earlier resigned but had not yet left Mexico, was mentioned in the *Mexico Daily* in an article calling for his expulsion from Mexico for his treatment of the nationals.[179]Again, this was not precipitated by William.

The unrest among the missionaries rose to such fever pitch that some openly talked to Mexican workers about their desire to see William expelled and to bring charges against Willingham and the home mission board for not firing him.[180]

On May 27, 1898 William wrote to Willingham:

> Accept the resignation of every one of us and use only native help. I am ready for anything, only you must not suffer for envy against me.[181]

On June 2, 1898 William wrote to Willingham saying that recently he startled even his physician when he was dangerously ill and had almost no pulse. However, William, the optimist, contended that he was better.

Soon after that William communicated to Richmond that the Home Board in Cuba had offered him an appointment. He wrote:

> I know the danger to me of a malarial country yet will do anything to get out of this muddle.
>
> I have not the least conception what complaints the brethren have against me at this time.[182]

Willingham immediately called William to Richmond but ended up accepting his resignation. Before William could move his family out of Mexico he had to be hospitalized twice more for various physical ailments. The second affliction caused rheumatism to set into both of his legs rendering him barely able to walk. On that occasion five doctors were called in to attend to him.

William was eventually cleared by the board of all charges brought against him. All three of the original accusing missionaries were gone from Mexico by the end of 1898. William bore no malice and continued helping arrange favors for the remaining missionaries, even those who had told him to resign.

William's reference to mistreatment in Mexico by other missionaries is only seen in his personal correspondence to Richmond at the time. I found no mention of it in his other writings.

The Southern Baptist archives indicate that for many years William continued his correspondence with Willingham. In one of those letters written from Kentucky in 1909 William indicated that he was at that time confined to his bedroom with malaria. He mentioned Willingham's daughter's engagement and also Florence's (my grandmother) impending engagement and stated:

> This matter of falling in love like measles, seems to be catching. Just a short time since [William's daughter] Mammie… became the wife of a fine young minister.… Another of our girls [Florence], who is a student at the Training School is holding some very suspicious interviews with her Mother each time she returns home, and I should not be surprised to hear that after next June, we will have only one daughter left. By the way, I am very proud of the high grade she is making in her examinations at the Seminary.[183]

Years after leaving Mexico William stated:

> In the national palace and in different state houses, I was often requested to explain to jurists and legislators, our [Baptist] love for democracy and the right of each individual to approach the Supreme Being without any mediator save our Lord Jesus Christ.
>
> President [of Mexico] Diaz would do anything for me which I asked.…I went to him with the troubles of Baptists, Methodists, Presbyterians, and Adventists, and he readily granted my requests.[184]

An Evangelical Saga quoted Alejandro Trevino, prestigious Mexican pastor, who wrote about his co-worker William Powell:

> "He was an untiring missionary. Gifted with vigor and strength he was able to ride horseback through the mountains of Mexico or travel by train to the USA to further the needs of the work…he owned friends everywhere. He was a magnet, a live wire. He was always looking for a new field or a new worker.
>
> "If a young person revealed gifts for the work, he immediately recruited him/her…he was the friend of governors and public officials…he became a friend of President [of Mexico] Porfirio Diaz…through these contacts he was useful to many missionaries of all denominations by acquiring freedoms denied by fanatical local officials."[185]

A History of Baptists in Southern States East of Mississippi (1898) stated:

> The most fruitful and progressive department of work under the Foreign Mission Board is that of the Mexican Mission.
>
> ....The qualities of leadership possessed by W. D. Powell made him the acknowledged director of the Mexican Mission. Wise in conception, resolute of purpose, courageous in execution, irresistible in energy, and yet gentle in disposition and consecrated at heart – Powell combines all the elements of a great missionary leader in a region like Mexico.
>
> ....In the adobe hut of the lowly Mexican, upon the remote ranch, in the crowded mart, before the frenzied mob, in the presence of the highest officers of State, or in the most cultured assemblage – he is equally the master of the situation. Fired with consecrated earnestness, he sways the Mexican mind...[186]

In a paper given at the Fourth International Conference on Baptist Studies in 2006 in Nova Scotia (over 100 years after the Powells left Mexico), Justice Anderson stated:

> Perhaps the most productive of all FMBSBC [Foreign Mission Board Southern Baptist Convention] missionaries in Mexico, was W. D. Powell.[187]

Cuba

After his time in Mexico, William was superintendent of Baptist mission work in Cuba for a year (1898 – 1899). He stated:

> My next field of labor was in Cuba...having labored in Mexico so long I was particularly fitted for this work...After a few months of preaching there was an average of one hundred conversions a month; but it seemed wrong to keep my family of ten in that unhealthy climate. For this reason I was forced to give up the work and return to the United States.[188]

Back to Tennessee and Union University

William stated:

> From Cuba I migrated to Tennessee locating in Jackson where I labored with Union University and did some pastoral work. The main building of this institution [Powell Chapel] bore my name and was destroyed by fire.[189]

Three different buildings at Union University were named after W. D. Powell. The first called "Powell Chapel" (not the same as Powell's Chapel Baptist Church) was completed in 1899.[190] It was named after him in honor of his work in Mexico and in thanks for his fundraising on behalf of the university.

Large building in background of this photo was Powell Chapel before it burned down.[191]
Photo was taken in 1904.

That two story red brick building called Powell Chapel was actually much more than a place of worship. On the main level was the auditorium and president's office. On the second floor were meeting rooms for fraternities and literary societies, and in the basement a gymnasium. The building also included a band room.[192] It burned down in 1912[193] but was soon rebuilt bearing the same name.

The Union University campus moved to its current location in 1975 where today the theatre carries W. D. Powell's name. The chapel at the current campus is appropriately named after his Union University classmate and friend – later the president of Union – George Savage.[194]

Kentucky

In approximately 1903 William went to Kentucky.[195] He stated:

> Then followed ten years as State Missionary Secretary for the Baptists of Kentucky. My labors in this capacity were very fruitful.
>
> There was a great increase in contributions to missions....Five hundred and twenty church buildings were erected or renovated by the aid of the state mission board.[196]

It was during William's time in Kentucky that a book he co-authored, *Primacy of State Missions,* was published, and "every copy sold without advertising."[197] (From William to present, are four generations of authors.)

Primacy of State Missions

COMMITTEE ON COMPILATION:

W. D. POWELL, D. D., of Kentucky
J. W. GILLON, D. D., of Tennessee
JOHN T. CHRISTIAN, D. D., of Arkansas

Price { Paper, 25 Cents
Cloth, 35 Cents

PUBLISHED BY THE
STATE SECRETARIES OF THE
SOUTHERN BAPTIST CONVENTION

Copyright 1912, By State Secretaries of the Southern Baptist Convention.

Primacy of State Missions co-cuathored by William D. Powell.

Field Secretary for Foreign Missions Board

From 1913 until his passing in 1934, William was a Field Secretary for the Foreign Mission Board of the Southern Baptist Convention.[199] His re-hiring by the Foreign Mission Board was proof positive that he had been cleared of all charges from his time in Mexico.

W. D. Powell.

As Secretary, William was known for his fundraising abilities – raising money for foreign missions, as well as for the paying off of the entire outstanding debt of many of the 734 churches that he dedicated during his 60 years as a preacher.[200]

Who's Who in America

Most of his fundraising was small donations from average people. However, according to the Luginbuel Funeral Home report he promoted literacy and education in the South and received funds for libraries from Andrew Carnegie and John D. Rockefeller.[201] William's prominence was evident in his several years of listings in *Who's Who in America.*[202]

The Final Years

William's granddaughter, Cita Harris Strunk, lived with her grandparents briefly when she was was young and her mother, Florence Powell Harris, had returned to college. At that time her grandmother taught her hymns. Cita recalls her grandfather as a tender and affectionate man. When he went on trips, he sent letters to 3-year-old Cita on stationery decorated with little elephants. "Many people regarded him as their best friend," she stated.[203]

W. D. Powell in Kentucky in early 20th century on Old Maude.[198]

W. D. Powell in his later years.

William Powell wished to keep preaching his whole life, and according to Cita he nearly achieved that goal – until he fell very ill over the last six months of his life.

The preface to *A Century of Baptist Work in Mexico* begins with author Frank Patterson's recollections from almost 35 years earlier of a powerful chapel message that William gave when he was past 70 years old. Patterson stated:

> The dynamic advocate of foreign missions who spoke in chapel was Dr. William D. Powell, Southern Baptists' first [long-term] missionary appointee to Mexico. Past seventy, his enthusiasm for Mexico was still contagious.[204]

After hearing Powell, Patterson was inspired to join a student group which conducted services in a Mexican community in the US. Patterson then went on to serve over 30 years with the Baptist Spanish Publishing House.[205]

William died at his daughter's home in Opelika, Alabama[206] on May 15, 1934.[207] His wife passed away only a few days later on June 2, 1934.[208] They had been married for over 58 years.

William's full page obituary in the *Western Recorder* stated that he was:

> Brilliant in mind, jovial in spirit, quick of wit, he readily won his way into the friendship of men and women of his denomination. He was very much in demand among pastors...[209]

Books that mention W. D. Powell but which I have not cited include: *A Global Introduction to Baptist Churches*, Robert E. Johnson, (2010); *A Texas Baptist History Sourcebook,* Joseph Early (2004); *Texas Baptists: A Sesquicentennial History,* Harry McBeth (1998); *The Baptist Heritage: Four Centuries of Baptist Witness*, H. Leon McBeth, (1987); *A Blossoming Desert: A Concise History of Texas Baptists*, Robert Baker (1970); *Frontiersmen of the Faith: A History of Baptist Pioneer Work in Texas,* Zane Mason (1970), and *A History of Baptists of Texas,* B.F. Riley (1907).

In an article published in 1930, four years before he died, William rejoiced that he had ten living children, thirty-five grandchildren, and one great-grandchild.[210]

The Bible teaches that Christianity is not inherited but is the conscious choice of each individual. Two other generations of foreign missionaries followed William and Florence Powell as well as several other individuals who served or continue to serve in various aspects of Christian ministry.

Three of the Powells' daughters married preachers.[211] One of those three, my grandmother Florence Powell Harris, along with her husband, served in China under the Southern Baptist Foreign Mission Board.

To date numerous individuals in at least six generations in our family have placed their faith in Christ—partly as the result of the slave lady who prayed.

The Second Generation: Missionaries to China

Hendon Mason Harris Sr. and Florence Powell Harris

William and Florence Powell had twelve children but two died in infancy.[212] Their seventh living child, also named Florence, was born in 1889 in Saltillo, Mexico.

The older Powell children were sent back to the United States one at a time to go to school. In the States those siblings lived with various friends and relatives. Therefore, much of the time they were not able to be raised together. The remaining family members, along with Florence, left Mexico when she was 10. Later she studied to prepare for missions.

Like her father, Florence was also a good story teller, intelligent, and dedicated to her calling. As her father William Powell had predicted in his 1909 letter to his friend and former supervisor, Dr. Willingham, Florence married in 1910.

Florence Powell age 8.

Hendon Harris and Florence wed on June 1, 1910 in Louisville, Kentucky. Two days later they visited the Foreign Mission Board in Richmond, Virginia. In November 1910, only days after Florence turned 20 years old and five months after their wedding, Hendon and Florence sailed to China as Southern Baptist missionaries.

As Florence's parents had done in Mexico, the young Harris couple purposely chose what they knew would be a difficult station. Hendon and Florence chose Kaifeng, Henan Province, in interior China.

Florence's husband, Hendon Harris, was born in 1885 near Jackson, Mississippi on what had once been the wealthy plantation of his maternal grandparents. Hendon's great grandfather, Dr. W. D. Hendon had purchased the estate as a wedding gift for daughter, Louisa, a graduate of Judson College.

Further back, his direct ancestors included a line of Scottish kings (the Stewarts), fiery Protestant reformer John Knox, the Witherspoons, and several men who fought in the American Revolution. Another of Hendon's family lines descended from William the Conqueror, through the Henrys and evil Prince John. He was a distant cousin to the current British royal family. [213]

However, when Hendon was a child, his parents were neither wealthy nor prominent. The cotton gin and grist mills on the plantation were the main sources of family income. One fall the gin operator, busy with customers, did not notice the group of children playing in the loose cotton. In a careless instant on the part of young Hendon, his foot was caught in the cotton baler. Acting quickly an adult reversed the machine, saving the boy's life.

Gangrene soon set into the wound. The doctor repeatedly insisted that the foot be amputated in order to preserve Hendon's life. His mother loudly resisted. Her prayers and constant bathing of Hendon's foot saved the limb, his life, and his future career. Missing one foot, surely he would have never later gone to China.

Within a few years the cotton gin and grist mill, which stood next to each other, both burned down the same day. The family had no insurance to replace them. Economic depression followed the fire and unfortunately they lost the farm.

From an early age Hendon practiced oratory. He overheard adults say, "This boy sho is goin' to make a preacher."

Hendon Harris as a boy and as a young man.

Early Missionary Work in China

Robert Morrison from England is credited with being the first Protestant missionary to China. He arrived in 1807 and at first had difficulty even getting anyone to teach him Chinese because China's government forbid teaching their language to foreigners. He had difficulty gaining entry into China proper. He spent most of his efforts learning Chinese and then translating the Bible and a dictionary. "…he found that lending out his books and tracts afforded...the best response."[214]

Morrison deeply mourned his wife and a son who died in China. Morrison's last years were spent composing evangelical texts and teaching others to operate his printing presses so that his work would continue after he was gone.[215]

The Bible is central to any Christian's understanding of and relationship with God. Therefore, Morrison, though he personally saw few converts, laid essential groundwork for all other Protestant missionaries to Chinese.

Hudson Taylor and his China Inland Mission (CIM) were the first Protestants to penetrate interior China. As a faith based mission out of Britain, they had no guaranteed income, so were constantly in prayer for God's provision. Living very simply, they dressed like the Chinese of that era. They wore long robes and the male missionaries wore their hair in Chinese style queues. Interior China proved difficult, but Henan Province seemed impossible.

Initially Henan Province violently drove out all missionaries who tried to go in. *Henan: Fire and Blood* states that in 1888 while Taylor was on a speaking tour in Canada:

> When [Jonathan] Goforth expressed interest in working in Henan Province, [Hudson] Taylor told him: "Brother if you would enter Henan, you must go forward on your knees." [Indicating that only prayer could penetrate that province.][216]

China Inland Mission was the first to serve in Henan Province. Goforth, under a Canadian mission, soon followed.

In 1900 while the Goforth family was mourning the recent death of their oldest child from meningitis, they were attacked as they tried to flee Boxer Rebels near Xintian. *Henan: Fire and Blood* states:

> The backs of the donkeys were broken to prevent escape, and then Goforth was struck down with the blunt edge of a sword, nearly breaking his neck… his arm was slashed to the bone in several places. When he struggled to his feet, he was struck unconscious with a club.[217]

Miraculously Goforth and his family received help from Muslim villagers, "fearful of …God's wrath if they had taken part in slaughtering the missionaries."[218] Later Goforth became known as "China's Greatest Evangelist."[219]

Within the most difficult province in China, Kaifeng was probably the most demanding station that Hendon and Florence, my grandparents, could have chosen. According to 1913 *Annual* of the Southern Baptist Convention: "Kaifeng was the last provincial capital in China to be opened to the Gospel."[220]

Henan: Fire and Blood states:

> In the late 1800's even Chinese Christians were not permitted to enter its [Kaifeng's] gates and were severely punished if they attempted to do so.[221]

Numerous times over the years the Chinese themselves, often at great risk, were the ones who spread the Gospel in China. The first known Christian in Kaifeng was a man named Zhu who acquired portions of Scripture and tracts from a Chinese vendor who sold books in 1881 outside Kaifeng city walls. When Zhu received these he was threatened by another man; the vendor's books were knocked into the dirt; and the salesman was beaten.

Through the reading of these Scriptures and then going to another town to visit a missionary, Zhu came to faith and led others to Christ. Therefore, there were already a few believers in Kaifeng before the first foreign missionaries were able to enter.[222]

In 1902 China Inland Mission (CIM) was finally able to rent a house in Kaifeng and began evangelistic activity. CIM was able to secure that location because the Chinese believed that house to be haunted.[223]

Fortunately for those in Kaifeng, Dr. Whitfield Guinness, initially one of only two missionary physicians for thousands of square miles, arrived there in 1902 to serve with the CIM. Guinness opened a hospital and trained Chinese doctors in Western medicine.

Earlier during the Boxer uprising Dr. Guinness, along with three other missionaries and a baby, narrowly escaped death. For 16 days they hid in a shelter surrounded by rebels seeking to kill them. During that entire time they all prayed that the baby would not cry – thus revealing their hiding place. That miraculous delivery changed Guinness's life. He became an ardent evangelist as well as a physician – leading many of his patients to saving faith in Christ.[224]

Hendon and Florence Enter China

Southern Baptists formed their foreign missions board in 1845 and their first foreign mission field was China.[225] However, they did not enter Kaifeng until 1908. By choosing Kaifeng only two years later, the young Harris couple was entering a very new and uncertain field.

The Hendon Harris Sr. family arrived in China 1910 and left for the last time in December 1947. During their tenure they experienced many victories. However, they also experienced anti-foreign uprisings, robberies of their home, and the deaths of several of their fellow workers, both Chinese and Caucasian – including that of their own infant son.[226] They also witnessed much human suffering brought on by wars, floods, famines, and epidemics. Despite these tragedies, after furloughs they repeatedly fought to re-enter China in order to serve there.

On Hendon and Florence's initial arrival, areas in China beyond the big cities and ports, including Henan Province, were ruled by roving armies of warlords.

China: A History states:

> The first half of the twentieth century saw China submerged in a "Period of Disunion" every bit as blood-soaked and confused…as the inter-dynastic free-for-alls of the past…. From 1911 to 1950 the fighting never really stopped; foreign invasion, became freedom struggle, became civil war, became revolution.[227]
>
> There were literally thousands of warlords, ranging from generals and officials who ruled whole provinces...to local toughs with a few hundred "braves," or restive hill minorities with an assertive chief.[228]

Of course, Hendon and Florence on their arrival could not have known what was ahead. However, they were well aware that only 10 years earlier, during the Boxer Rebellion of 1900, many Chinese Christians, foreign missionaries, and their families had been attacked, raped, tortured, or executed in China. The Boxer rebels destroyed railroads and telegraph facilities, thus hindering exit.

China: A History states regarding the 1900 Boxer Rebellion:"Trouble rapidly spread through Hebei, Shanxi and part of Henan, where many foreigners, mostly missionaries, were massacred."[229]

According to *Henan: Fire and Blood:*

> Not only foreigners were targeted but also those Chinese known to be associated with them. Thousands of native Christians – both Protestant and Catholic – were slaughtered, as were more than 300 foreign missionaries.[230]

Southern Baptist missionary Annie Sallee discussed the effect of the Boxer Rebellion:

> Before the blood had ceased flowing, and before the last missionary had been called "to suffer dishonor for the Name," the news spread to Christian lands, and there were more volunteers than ever before, not only for China but for other difficult fields. Thus the "blood of the martyrs" once more proved to be "the seed of the church."[231]

On arriving in China the Harris couple visited Kaifeng briefly, but their first year was mainly spent in language study in Zhengzhou, Henan Province – about 50 miles away. The Harris's first child, Miriam, arrived March 31, 1911, while the family still lived there. Florence wrote:

> Life in Cheng Chow [now spelled Zhengzhou] was quite different from that in Kaifeng, for here the homes and institutions of the Southern Baptists were scattered over several acres of wide open country with trees, flowers,

and gardens around them. Here were a hospital, a doctor, and an American nurse and two interesting families – the Lawtons and Herrings.

Mr. and Mrs. Herring were outstanding in their ability to speak and preach in Chinese. Pretty Mrs. Herring, a very spiritually-minded lady and an accomplished musician, was an ornament in our missionary family.

Mr. and Mrs. W. W. Lawton, so affable and hospitable, made a contribution of their own. Mr. Lawton had a special gift for smoothing out the rough places for both Chinese and Americans. He was one of the kindliest, most thoughtful persons I ever knew.[232]

Kaifeng

After language studies, which both Hendon and Florence completed, the Harris family settled in Kaifeng, Henan Province[233]only six miles from the Huang He (Yellow River). The literal translation of "Henan" is "south of the river."

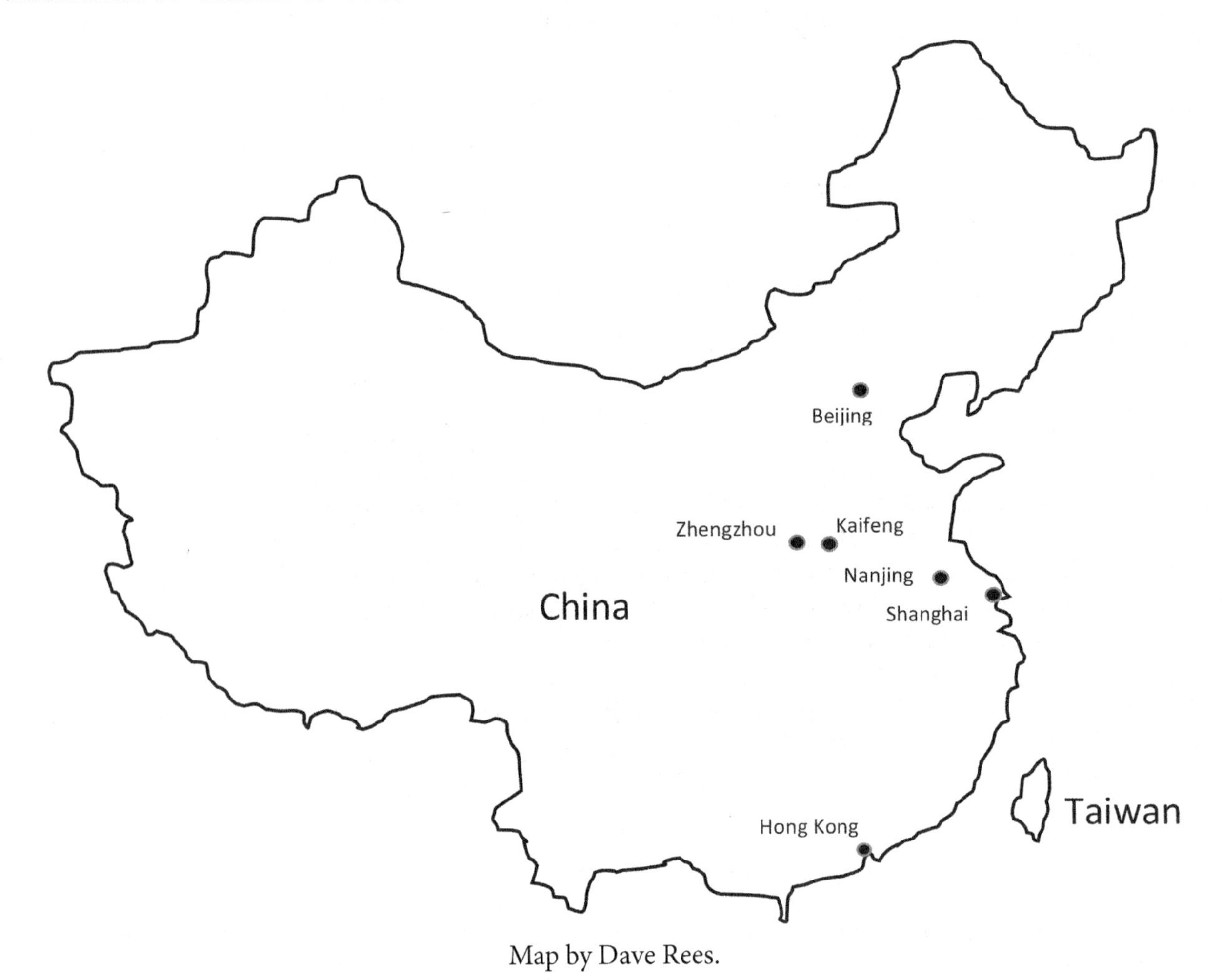

Map by Dave Rees.

The main source of livelihood in that region was farming. Irrigation from the Yellow River was and is the lifeblood of that dry, sandy area. However, the river frequently flooded, including in Kaifeng. Several times those floods killed thousands.

At that time Kaifeng was enclosed by a high wall thirteen miles around with six gates. During the Sung Dynasty (960-1127), Kaifeng had been the capital of China. Once a terminus of the Silk Road, camels still frequented the town. Kaifeng was known for hot summers, cold winters, and sand storms.

James Gillespie, son of missionaries, recalled one of those sand storms:

> One day we were outside and our Chinese amah [nanny]...began to shout and point to the north. The sky was a dark yellow-brown color. They grabbed us up and rushed us into the house. We found ourselves in the midst of a terrible dust storm. Everyone was frantic...The wind blew and blew.
>
> The dust sifted in through the window sills. Efforts to seal the windows with newspapers and household glue failed. The house was filled with the finest particles of dust before the storm stopped blowing. The dust storm had blown down from the Gobi Desert in Mongolia...Those storms were... not unlike those in the Western United States in the 1930's.[234]

Kaifeng is also famous for its Jewish community. *Journey into China* stated: "The first Jews in Kaifeng – probably traders – founded a community...They built the first synagogue in 1163."[235]

The Kaifeng Jews have intermarried so look Chinese, but some of them continue their Jewish traditions. The cover of my grandmother's book, *How Beautiful the Feet,* sports a photo of my grandfather and two of the Kaifeng Jews.

The story goes that a few years back Jews from Israel traveled to visit the Kaifeng Jews, who took one look at those from Israel and said, "But you don't look Jewish!"

There is controversy about whether Marco Polo actually reached China. Some state that he arrived in Kaifeng in 1265,[236] more than a hundred years after the Kaifeng Jews built their first synagogue there. Ironically Polo's writings gave the startling allegation that he saw Jews in China.

Florence stated that during the time the Harris family lived in Kaifeng, the city of 300,000 had neither water nor sewage systems. Water had to be carried in and sewage carried out daily. Laborers, each with two open wooden buckets balanced on a pole over his shoulder, took the human manure out of the city to farmers who used it to fertilize crops.

Southern Baptists, W. Eugene and Annie Sallee, had moved to Kaifeng in 1908. It was a difficult station, but there was plenty to do for everyone and reportedly sweet fellowship between the missionaries of the different agencies.[237] Later cooperation of all the missionaries during the Japanese invasion helped preserve many Chinese lives and saved properties from Japanese occupation.

How Beautiful The Feet

By

Florence Powell Harris

By the time the Sallees left on furlough in 1911 they had already started separate boarding schools for boys and girls and a church in Kaifeng. Their departure left in place teachers for the schools, a Chinese evangelist in charge of the church of 200, and a "Bible woman" for women's ministries.[238] When the Sallees returned, they continued to administer the schools. For many years Sallee was the senior Southern Baptist missionary for that station.

Education of Women in China

Hendon wrote that it was missions that initially elevated the status of women in China: "Until mission schools were opened for girls, Chinese women had practically no educational opportunities."[239]

Today Chinese universities enroll both men and women, so it is difficult to envision that lack of education for females in China 100 years ago. However, according to John Keay in *China: A History*: "As late as 1909 only about 13,000 girls were enrolled in schools (all run by missions) in the whole of China…"[240]

Charles Maddry, Executive Secretary of the Foreign Mission Board of the Southern Baptists, wrote in *Christ's Expendables* that later:

> The [Baptist] University of Shanghai was the first in all China to open its doors to women on terms of absolute equality with the men….It was given to the Baptists with their passion for democracy and the equality of every soul before God, to lead in coeducation in China.[241]

In 1902 China had banned the cruel practice of foot binding, but individuals in many outlying areas resisted. Foot binding had been practiced in China since the 11th century. The ideal (but seldom achieved) length of the foot was three inches.

Foot-binding required starting at a young age and repeatedly breaking the girl's foot bones. Each time the foot and toes were folded under. That led to much pain, and sometimes even death – should infection set in. At that time many Chinese men felt that bound feet made women look beautiful. Parents put their daughters through that ordeal to make them marriageable.

In truth, foot-binding subjugated females to pain and captivity. It would have been difficult to run away with such feet. Once the feet were broken and misshaped they were almost impossible to correct.

Florence, who was petite with proportionate feet, reported being mocked by Chinese men who deemed her unbound feet big and ugly. The title of her book *How Beautiful the Feet,* is taken from Isaiah 52:7: "How beautiful on the mountains are the feet of those who bring good news, who proclaim peace, who bring good tidings, who proclaim salvation…" However, I believe that the title of her tome is also a retort to the criticism of her feet.

China: A History states that Christian schools in that era usually required their female students to have unbound feet. It would have been difficult for schools to have dealt with the daily care of each child's bound feet. The requirement for the students to have unbound feet surely gave parents incentive to comply with the new law.

The Role of Southern Baptist Missionaries to China

Hendon wrote regarding the role of foreign missionaries:

> ...missionaries are the guests of China...the missionaries sacrifice to come and serve China and the Chinese without ulterior motives. Charges that they come with the desire to exploit the Chinese and debase their civilization, can only be the result of ignorance of the missionary enterprise. Missionaries cannot be expert in the realm of politics and therefore should not attempt to direct ...policies of governments: nevertheless, in cases such as the opium traffic, the coolie traffic and open and flagrant breaches of international morality, they should speak out as in times past.[242]
>
> There can be no doubt that in many kinds of activity in religious work, the Chinese have qualifications that missionaries lack and can never attain. It has often been said that China will be finally Christianized by the Chinese themselves. The present period demands intensified training for national pastors, teachers, church workers and other religious leaders. Consequently the missionaries should give themselves more and more to the task of training and advising those who are to come into the control.[243]

From the very earliest of the Southern Baptist time in Kaifeng, Chinese pastors and helpers served right alongside the missionaries. The first pastor of the Baptist Church in Kaifeng was Dr. Li Yung Ching,[244] who was there until his death. He was followed by another Chinese man. For many years Dr. Peter Lee was pastor of the Kulo Baptist church in Kaifeng.[245]

Life in Kaifeng

Life in Kaifeng in the early twentieth century was not easy. Florence wrote of her initial loneliness there. No sooner had the Harrises settled into Kaifeng when because of unrest between the Chinese and Manchus (people from Manchuria, an ethnic minority living in China) they had to temporarily flee to another area carrying infant Miriam.

The Chinese were threatening at that time to behead the Manchu men, women, and children – right outside the city wall near the missionary compound.[246] When the foreigners fled so did the students in the school within the compound.[247]

The 1912 *Annual* states:

> Mr. and Mrs. Harris had just moved into their new home...when the threatening conditions, caused by the revolution made them leave. They were in their new quarters less than one month.[248]

Revolution in China—beheaded civilian looters. Photos by Walter Stursberg published in London 1912.

After the Harrises returned to Kaifeng, they had wanted to live a simple life. However, after they were repeatedly robbed – once when thieves broke through the brick walls of their home at night while they were inside – they employed servants to guard and to help with household chores. After one robbery, Hendon chased the robbers and caught one by his long braid of hair, but he let go when the man wielded a knife. Baby Helen arrived in 1912.

Florence and Hendon with Helen and Miriam.

One China missionary report in the 1912 *Annual* of the Southern Baptist Convention stated:

> … native pastors and churches… take charge of their local home mission work. This is developing the native churches in responsibility and will be a great source of good. It will be a glorious day for the mission work in other lands when the natives can take charge of the work of sending the gospel to their own people.[249]

Hendon felt that trained Chinese were perfectly capable of running their own churches. Eventually he was able to begin a graduate level seminary which from the start had both Chinese and Americans on the faculty.

Over the years Hendon Harris had many responsibilities including relief work, but his main assignment was to be an evangelist and church planter which required much travel into the countryside. His doctoral thesis clearly states that his goal was not to make the Chinese into Americans but rather to tell them about Jesus.

Protestant evangelists in China taught that Shang Di, the supreme god described and worshipped by early Chinese rulers – the god that all Chinese knew about, was the same God that the missionaries worshipped. Shang Di was the Creator and Lord of all things.[250]

Their message went on to say that all people were sinners. Shang Di was concerned about lives of individuals but required a blood sacrifice for sin. The Bible taught the same. No one by his own merit could receive forgiveness. According to the Bible Jesus died on the cross to pay the blood sacrifice for our sins. He was buried but is so powerful that he came back to life.

Now Jesus lives in Heaven with Shang Di (the Heavenly Father) where he prays for us. The missionaries explained that in order for a person to receive the gift of forgiveness, he had to individually place his faith in Shang Di. The story of forgiveness and redemption was in the Bible. Since many of the people who heard this message were illiterate, the missionaries taught literacy so that the Chinese could read the Bible themselves.

Bible verses often used in their messages were:

Hendon in China on Donkey.

Romans 3:23 "For all have sinned and fall short of the glory of God."

I Corinthians 15:3 "…Christ died for our sins according to the Scriptures…He was buried, and…He rose again the third day…"

Romans 6:23 "For the wages of sin is death, but the gift of God is eternal life in Christ Jesus our Lord."

Ephesians 2:8 & 9: "For by grace you have been saved through faith and that not of yourselves; it is the gift of God, not of works, lest anyone should boast."[251]

Evangelism in the Villages

Florence stated that Hendon was so tall that when he traveled on a donkey for his evangelistic work, he was almost too large for the animal. Several years later, some benefactors in America sent him a model-T Ford. Even with a car, it usually took most of a day to cover 40 miles because the roads were rough dirt.

Florence wrote about Hendon's first outstation convert:[252]

> The crowd of on-lookers stayed entranced with this foreign stranger until nearly midnight and wouldn't leave until he blew out his candle....
>
> Early the next morning the visitors of the night before returned, bringing their neighbors and kinsmen to see the "monstrosity" [Hendon] and hear about America. As the room became over-crowded, the young missionary suggested…"How about us all moving out to the street, for I have come thousands of miles to tell you the most wonderful news." They had wondered why he had come.
>
> …he and they were soon joined by a host of journeymen, who were attracted by the large crowd around the white man. Now he had a large audience to preach to.
>
> Here in the noon-day sun stood this ambassador for Christ telling the story of Jesus for the first time to these hearers. He started off by referring to their many temple idols among which was not one who answered prayer. Then he added that Jesus was alive and had the power not only to answer prayers, but to change sinful hearts and make them in his image.
>
> He didn't get to finish his sermon, for a young man stepped out of the throng, saying, "Mister, I want my heart cleansed like that you are telling us about. Will you ask your God to give me that kind of heart?"
>
> Two other men came to join the first….the four of them knelt down in the deep dust of that road…and God heard that prayer. These three converts became the nucleus of the first church of Chang-shih.[253]
>
> …whenever a church was established, a Christian school was begun in the same courtyard. Cities were opened up to Christianity and churches organized faster than there were trained national leaders to man them.[254]

Fellow missionary Annie Sallee wrote: "Mr. Harris' work of evangelization was being felt in the city as well as in the districts surrounding Kaifeng."[255]

Florence recalled one time when a group from an outlying town, came to ask for help in building their church:

> Late one afternoon a band of thirty Christian men and women came to our home to interview Pastor Harris on an urgent request. The church group, including a number of foot-bound women, had trekked the thirty miles on foot from Pei Village to Kaifeng.
>
> The women, because of the "middle age spread" and their tortured feet, had to use walking canes to help them over the road. The men…walked slowly behind at a pace convenient to their women folk. They couldn't afford to

> hire a ride, not even on the lowly donkey...The people were so tired that they almost dropped into the chairs offered them in our reception hall.
>
> From another room I heard this conversation: "Pastor Harris, you do not know us, for we are a delegation of new Christians from Pei Village. The Gospel came to our town a short time ago and the thirty of us here accepted the Lord. We have been meeting in each other's homes for Sunday services...but now we long for a real church building.
>
> "Recently we called a church meeting to discuss a new house of worship, and this is what we found: There are masons and carpenters in our number who promise to erect the building. All of us farmers will contribute dirt for the brick we expect to make. (The average farm has from one to three acres.) Some of the members have trees they have offered for the woodwork, and all of us will help build the church.
>
> "Pastor, we are confronted with a serious problem. The thirty of us can't raise the amount...we need to buy the building lot, so we decided to come to see if you couldn't devise some plan to help us get this land."[256]

Although the Baptist missionaries frequently contributed of their own funds for famine relief and various other projects,[257] at that time the Harris couple did not have the financial means to offer. However, Florence, through relatives in a church in the US, quickly raised the amount needed to buy that lot. Pei Village became an outstanding Christian center. Years later Paul Pei, who grew up in that village, was an instructor at China Baptist Theological Seminary in Henan.[258]

Later an American philanthropist, a member of another denomination, sent $1000 a year during the last 12 years of his life to enable young Chinese ministers to attend school in another province and to help furnish the new churches and schools.

Pastor Chao, originally from Henan, was one of the first evangelists sent back to his home town after graduation. When Hendon started and became head of Henan-Anhuei Bible School, Pastor Chao was on the faculty. At that school both men and women received theological training in their own dialect while remaining closer to home.

The 1914 *Annual* (which described the mission work of 1912) contains a report from H. M. Harris (my grandfather):

> The earlier part of the year I gave most of my time to language study. Besides this, I preached a number of times in the city Chapel, and conducted services each Sunday in the Nankuan Chapel, also teaching a Sunday school class and attending the meetings of the Honan [now spelled Henan] Famine Relief Committee, of which I am secretary.
>
> The Relief Committee distributed several thousand dollars which had been forwarded from the Shanghai Relief Committee, and had been supplemented by gifts from Kaifeng and elsewhere.

> The famine was due to drought which destroyed the wheat crop over a large area, and in some districts the misery was aggravated by the depredations of large robber bands. Forty thousand hungry people were fed in Kaifeng by the government; scores died of starvation and malnutrition. Some ate ground rock and the bark of trees to allay the pangs of hunger.
>
> The money distributed was for the most part used to build roads, thus giving employment for those who were in enforced idleness. Indications are that we are to face a similar situation again this winter as the wheat crop has failed again, and in some sections the robbers roam at will.
>
> Out-stations. In the fall I began regular work at two out-stations, Chu Hsien Chen and Wei Si Hsien, fifteen and… thirty miles south of Kaifeng. I have as my assistants, Mr. Shih, a well-trained native evangelist, and a colporteur [peddler of religious books] for the important work of sowing broadcast Gospels, portions of Scripture, and tracts. The district is practically untouched by the gospel, only one Christian living in it.
>
> South Suburb. We have had preaching at the small chapel in the south suburb all the year. This district lies between the south gate of the city and the railway station, increasing daily in population and importance. This fall I conducted an inquirers class for two weeks, and following that had over a week's preaching at night, assisted by my native helper. In December a small day school was begun in this district.[259]

In Asian culture "saving face" is paramount. "Workfare" kept people from feeling like beggars. The roads were dirt, easily torn up by carts, and turned to deep mud when wet. They constantly needed repair. Years later, during the 1940's and the greatest relief efforts in Henan province, one missionary reported seeing people of every description working on roads outside their respective villages.[260]

Hendon and Florence's first son, William Powell Harris (named after Florence's father), arrived in late 1913, but he died at 13 months. The China field report of the Southern Baptists for that year stated that Pastor Li and two American missionaries died of typhus. That report also mentioned the death of the "Harris baby boy."[261]

The Worth of a Child

It is hard to imagine the anguish of William's parents at the loss of their infant. They were so young and so far from the support of family. Years later Florence wrote:

> William Powell was a happy, sunny-faced little boy who won all our hearts. We couldn't understand why this little fellow had to be taken from us when only thirteen months old. He died of pneumonia following a child's disease.
>
> His body was laid in a casket that was trimmed and cushioned inside with silk. This was the work of love of Mrs. W. E. Sallee. One non-Christian remarked, "Isn't it a terrible waste to put all that silk in the casket?" When

Chinese children die, they are stripped of clothing, then thrown in a field, or at best buried in a shallow grave that wild dogs can easily find.

He was laid to rest in the Cheng Chow Mission Cemetery. In this sad experience the sacredness of a child's body was exemplified.

About…1938 a Japanese bomb fell on…[William's] grave, completely obliterating it, even the head stone.[262]

Hendon wrote in his 1927 doctoral thesis, "Indigenous Churches in China:"

In Christian teaching, "everybody is somebody"….Children in the family have a value set on them by the teaching and example of Jesus…. When a baby dies in Honan Province, the old way has been to fling the little one away unburied to be eaten by dogs. The ordinary way to speak of the death of an infant is to say that is "rung la" (flung away)…. the truth is these words may come from a broken heart…there is fear that the devil that took away the child, may return if the mourning is too deep and continued. Christianity is giving the children a new dignity and worth in the eyes of the Chinese that they have not had before.[263]

In that era there were frequent famines. Sometimes desperate families abandoned children. Florence wrote:

Once Mr. Lawton [fellow missionary] came across a pitiful little beggar boy who was crying his heart out because his father had deserted him. This little waif was dirty, scabrous, and emaciated, and it was soon discovered that he was tubercular. Of course, the boy was hungry. Like the Good Samaritan, this missionary did not pass on by, but instead, took the child in, clothed, fed, and sent him to the hospital for treatment and later sent him to school.

Out of this damaged chrysalis emerged a frail, spirit-filled, lovable young man, Mu Rung Kwang, who later became the Young People's leader for the entire Interior Mission. He was a rare jewel. Had Mr. Lawton done nothing else in China except find and nurture Mu Rung Kwang it would have been worth his going to China.[264]

The Baptist Compound in Kaifeng

The 1915 *Annual* describes the Baptist compound in Kaifeng:

For evangelistic work in Kaifeng, the center is the city compound on the busy, well-known "Drum Tower" street. Here is the book room, street chapel, guest room, day school for small boys, day school for small girls, night school, women's industrial school, chapel for Sunday services and church gatherings.

Gate to Baptist compound.

Some of the missionary residences, including those of the Harris and Sallee families, as well as those of some Chinese church workers, were also in that compound which was surrounded by a high wall.[265] The wall was about six foot high.

Florence stated that she also had a vegetable garden with American varieties of vegetables behind her residence. She wrote that during her years in China, she never owned an icebox or a refrigerator – the well was used to cool butter and milk.

Harris residence in Baptist compound was owned by the mission so was occupied at various times by different families. Note Florence's roses at the front of the house.

The Prayed for Son

Every family in that era longed for a son to carry on the family name. The arrival of baby Hendon Harris Jr. in 1916 was an answer to Hendon and Florence's prayers following William's death.

The Chinese must have been especially impressed that Hendon Jr. – this "prayed for son" – was born in 1916, the year of the Dragon. According to Chinese beliefs "dragons" are born leaders. They are feisty, gifted, and help right wrongs – including bringing justice over injustice. The dragon was the symbol of the emperor. Every Chinese parent hoped for a dragon child.[266]

Helen, Florence, Hendon Jr., Hendon Sr., Miriam.

Bao in Chinese means "to hold or to hug." Hendon Jr. was nicknamed "*Bao Bao*" because as a young child he always wanted to be hugged or carried about. As an adult he fulfilled most of the myths about dragon children.

Trip Home to America

Miriam was six, Helen five, and Hendon Jr. was only thirteen months old in 1917 when the Harris family sailed for their first furlough to America. According to the Canadian Passenger Lists, they landed at Victoria, British Columbia on June 13, 1917. From there they travelled by train across the North American continent.

Hendon, Florence, and their children visited the Harris grandparents who lived on a busy street in Birmingham, Alabama. Hendon Jr. and his siblings were used to the confines of a compound in China, so were warned to "keep off the street lest they get run over."

Helen, Hendon Jr., and Miriam.

Florence wrote:

> Our very small son [Hendon Jr.] was so impressed by "getting run over" that he decided to try it. One day he stretched himself out on the concrete highway dressed in a faded...playsuit that blended perfectly in the background, making him almost invisible; and he was run over by a young man in a car. Only the body of the auto, not the tires, went over Hendon Jr.
>
> The driver was terrified. Stopping his car, he walked back and picked up the child, even shaking him gently to see if he was unhurt. Still unconvinced, he took Hendon to a doctor for an examination. He was so relieved on finding the child unhurt that he bought his "victim" a bag of candy and drove him back to his grandmother's house.[267]

At that time furloughs normally lasted one year, but that trip to America stretched into two years – probably because of World War I. During the first year Hendon Sr. filled in for a pastor in Clinton, Mississippi.

Addie Cox

During 1917 Hendon was invited to speak at the Alabama State Convention. There he met a devout young woman, Miss Addie E. Cox, who wanted to go as a missionary to China but was discouraged because her support was not coming in.

Florence recalled:

> Hendon, impressed with her [Addie Cox], button-holed the state leaders in her behalf and made an appeal for her missionary salary before the convention, which was granted. Addie became Alabama's outstanding missionary daughter.[268]

Called to China stated:

> Women were essential to the success of the gospel in China....Officially [Chinese] women were discouraged from traveling about in public [or] conversing with males.[269]

Ministering as an evangelist among the women and children in the same areas where Hendon worked with the men, Addie Cox was tireless. *Biographical Dictionary of Chinese Christianity* states regarding Cox:

> She visited hundreds of villages each year, traveling by foot or riding donkeys, camels, wheelbarrows, carts, or bicycles, preaching to the crowds gathered in villages through which she passed. One fall she spent sixteen weeks on a single trip, returned home to rest, then left for another three weeks.[270]
>
> Addie taught the Bible, taught women to read, started a prison ministry for women, was involved in refugee ministries, and founded a school for women that taught spinning, weaving, shoe making, and other crafts.[271]

The Chinese Labor Corps in France

After Hendon's 1917 year as a pastor in Mississippi, the family remained there for another year while Hendon served in France and Belgium in order to work with the Chinese Labor Corps. Hendon wrote that 140 thousand Chinese had been sent to Europe in that effort.[272]

In World War I the British, with cooperation of Chinese officials, recruited thousands of young Chinese men for the Chinese Labor Corps (CLC) to do manual labor for the war in France. The recruitment was controversial; the Chinese workers were not treated with dignity; and reportedly they were purposely denied translators. It was a recipe for disaster.

Charles Hayford wrote concerning the CLC in France:

> By January 1917, the British were able to send the first of some 120,000 Chinese workers, and the French approximately 30,000. Mostly from Shantung [Shandong Province], they were shipped to Vancouver, then overland through Canada in what were reputed to be sealed freight cars, and finally across the Atlantic.
>
> ....A near riot broke out when a British officer started the day's march by shouting, "Let's go!!" a command which sounds in Chinese like the word "kou," [gou] or "dog." The workers assumed they had been basely insulted and refused to move.
>
> Between June 1917 and June 1918, ten CLC workers were executed for infractions of discipline and over sixty were combat casualties. There were more than thirteen strikes, six cases of group insubordination, and three of mutiny.
>
> By 1918 the morale of the British CLC workers had reached a low, almost desperate point. Earlier the British army command had resisted Chinese-speaking Y [YMCA] secretaries coming into the camps, fearing that sympathy would dilute discipline....in March of 1918 the high command invited the British Y to take over the canteens and recreational work ...G. H. Cole came from China...other veteran China hands came to assist him... Through the Y network among American colleges, over forty Chinese students and about twenty China missionaries answered the call.[273]

Hendon had nothing to do with the recruitment of the CLC. Those soldiers were even from a different province from his. Southern Baptist Biographical Data regarding Hendon states:

> ...his heart was stirred by the needs of the soldiers of the Chinese army in France....tho [sic] it was a difficult decision to make, he left his wife and children...and into the conflict he went.[274]

Because Hendon spoke Mandarin fluently and understood the Chinese viewpoint, he was able to intercede for the Chinese workers with the British officers – even at one point saving a condemned Chinese man from what would have been desertion charges and execution.[275]

In Europe Harris officially worked for the YMCA, but he was billeted in the British Army Headquarters. His work with the Chinese also allowed him to lecture about the Bible.[276]

In Sickness and in Health

In the fall of 1918 while Hendon was in France, Florence and her four children (Eugene was a newborn.) came down simultaneously with the raging, virulent flu that took the lives of so many people around the world that year. According to a report by Stanford University:

> The influenza pandemic of 1918-1919 killed more people than…World War I at somewhere between 20 and 40 million people. It has been cited as the most devastating epidemic in recorded world history. More people died of influenza in a single year than in four years of the Black Death Bubonic Plague from 1347 to 1351. Known as "Spanish Flu" or "La Grippe" the influenza of 1918-1919 was a global disaster… an estimated 675,000 Americans died of influenza during the pandemic.[277]

In order to care for the children while sick herself, Florence and the four children slept in two double beds in Florence's bedroom. With help from neighbors and her pastor, who daily checked on them, Florence was able to nurse all of them back to health.

Florence stated: "When Hendon [Jr.] was three he preached his first sermon on the sun, the moon and the stars."[278] She also stated:

> …one spring afternoon my little son, Hendon, who was three, climbed over the fence into the…[neighbor's] pasture and drove the entire herd, including the large bull, up to our dividing line. It was a dangerous feat, and hard on the cows to be driven up at three in the afternoon. So, to impress these facts on him, I met him with a switch in hand.
>
> Looking up at me innocently with his big blue eyes, this miniature herdsman pleaded, "You wouldn't whip a good lil' boy, would you, Mother?"
>
> "No, I won't this time, if you promise not ever to do this again."[279]

During his stint in France, Hendon Sr. was offered a lucrative position as a US deportation officer in Seattle but chose instead to return to China as a missionary.

The Harris Family Returns to China

Late in 1919 the Harris family sailed back to China. Baby Eugene, who had been only two months old and deathly ill the year earlier when Florence and the children had flu, contracted pneumonia during that trip across the Pacific. The other young Harris children shared another cabin where the ship stewardess had charge of them.

Hendon Jr. age 3 was delighted with his Christmas gift – a small wagon. Another passenger, who had not made it home for Christmas with her own children, ridiculed that Florence's children only received one present each for Christmas.

On New Year's Eve the ship stayed docked in Kobe, Japan for several days so the crew could celebrate. However, that first night in port the ship's baker was stabbed in the hallway near the children's room. According to Florence, for several days the children had to step over his corpse to get to their cabin.

During that trip Hendon Sr. contracted malaria followed by bronchial pneumonia. On reaching Kaifeng he was hospitalized. "For weeks his life hung in the balance" as he hemorrhaged from his lungs.

The Young Harris Children in China

The next Christmas, Hendon Jr. age four was again delighted with his Christmas gift – a tin horn. Addie Cox, on passing his home, saw him outside blowing his horn while looking up. "What are you doing?" she asked. He replied that he was having so much fun that he wanted the angels and Jesus to hear him. Years later his sister Miriam wrote a poem:

"The Horn"

By Miriam Harris Mills

Long years ago on a Christmas morn
A little boy found a horn
In his stocking, and laughed with glee
The shiny, pretty toy to see.
A few late stars were almost spent
When he in night clothes outdoors went.
The snow lay cold upon the ground,
But God's warm love did him surround;
He raised his horn and a few notes blew -
Oh, Jesus, thanks, oh thanks to you.
The years have passed, his hair is grey'd
On mission fields his life is stayed;
Still every day in dawn's chill dew
His heart is as boy anew, and cries,
Oh, Jesus, I still thank you.[280]

Hendon Jr. was born in 1916 and his three brothers followed in close succession – Eugene Sallee Harris (named after Hendon Sr.'s close missionary colleague) in 1918, Lawrence in 1920, and Richard in 1922. In 1926 baby sister Florencita (Cita) made her arrival.

Since her parents also had many children, Florence was not alarmed about having such a large family. She joked that her boys were not stair steps but rather a slide.

Hendon (5), Eugene (3), Lawrence (1) in Kaifeng.

The Harris children learned Chinese as early as they learned English. They had few toys, but they usually had books. Florence wrote that since children's playthings were not often available for purchase in Kaifeng, she would sometimes sketch for the local artisans who would turn the designs into toys for her children.

For most of their time in China the Harris children were kept within the missionary compound because it caused such a stir when the nationals saw the foreign children. To the Chinese they were peculiar looking. Florence wrote about being followed and taunted when she took the children out. As one racial slur some Chinese called Caucasians *gui,* which in Chinese means "devil or ghost."

On one occasion young Hendon Jr. overheard the household servants talking about a neighbor who was on his death bed. Wanting to see for himself, Hendon Jr. slipped out and pressed his face against the man's window. Young Hendon was very blond and fair. The superstitious dying Chinese man and on seeing such a strange, small, white being mistook him for a *gui* who had come to claim his soul. Sitting straight up in bed, the man screamed, frightening both himself and the inquisitive child outside.

Henan: Fire and Blood stated that banditry and disease increased in the 1920s along with civil war. The 1920's and 30's were times of great danger in Henan.[281] Annie Sallee concurred:

> Disgruntled citizens and those robbed of their flocks and homes, and some even robbed of their wives and daughters, have joined the great hordes of bandits which have made…Honan their lair, from which they swooped down on villages to pillage.

Annie Sallee then quoted her husband:

> "There has scarcely been a month…in which our province has been free from fighting….It would be difficult to say which the people dread more, the soldiers or the bandits."[282]

No Foreign Bones in China by World War II correspondent Peter Stursberg discusses China of that era. Not long ago I learned that Stursberg, who lived in Kaifeng where his father had been sent by the British to be the postal commissioner, was a childhood friend of my father. Kaifeng was so far inland that according to Stursberg family diaries, in October 1920 it took them seven days to reach Kaifeng from the port city of Shanghai.

Kaifeng Through a Child's Eyes

In Kaifeng even children sheltered most of the time in compounds could not avoid seeing atrocities. Stursberg wrote:

> The heads of recently executed criminals and so-called bandits were hung near the main South Gate and could be seen by my brother and me as we drove in our open carriage into the city, despite my mother's best efforts to divert our attention. My friend Miriam Harris, the daughter of a Southern Baptist missionary, actually witnessed an execution, even though her mother quickly covered her head with the carriage rug….There were beggars and lepers on the streets.[283]

In the early twentieth century during times of civil unrest, Pastor Hendon Harris sometimes was called upon to act as an intermediary between warring groups.

Several missionary biographies, including Florence's, discuss the shocking arrival to Kaifeng in 1922 of Feng Yu Xiang, "The Christian General."[284]

Stursberg confirms an account in my grandmothers' book about when Chao-Ti, the warlord in charge of Kaifeng at the time, was being chased out by Feng Yu Xiang. Hendon Harris was summoned to the palace of the reigning warlord and asked to negotiate for amnesty. Chao-Ti named Hendon head of a group of men ordered to seek amnesty with Feng Yu Xiang.

A truce team was sent to try to negotiate peace with the incoming warlord. Harris is the tall man in the photo.

Stursberg (whom in 2010 I was finally able to track down and meet in person at his home in British Columbia) gave permission for this lengthy quote below from his *No Foreign Bones in China:*

> Reverend Hendon Harris, a leading Southern Baptist missionary…was the father of my friends, Miriam and Hendon Jr.; he was a tall, good-looking Southerner….It would be a dangerous undertaking, but Mr. Harris, who spoke Chinese fluently, could not refuse. The commission was given a train – an engine on which large white flags were mounted, a coach, and a dining car – to take them to the border town of Chengchow, a railway junction, that the…[approaching war lord] had seized.
>
> Despite the white flags, the train was fired on as it approached Chengchow. Bullets blasted through the coaches, and the missionary and others …. dived for the floor. The train stopped. A courier, who, Mr. Harris told Dad, was one of his Post men, bravely went forward on foot with the white flag to explain that this was a truce mission.
>
> The train was allowed to proceed to Chengchow, where Hendon Harris had his first encounter with the famous warlord. He found him to be a huge man by Chinese standards, six feet, two inches tall, which was about the American's own height, but heavier; a charismatic figure, dressed in the

cotton uniform of an ordinary soldier. When Harris put forward Chao-Ti's plea, General Feng spoke directly without any convoluted courtesies: "No! No compromise! We are marching on Kaifeng!"

It took two days for the army to reach the provincial capital, and they were two days of chaos and anarchy, a classic example of what happens when the ruler has fled and all order and restraints have collapsed.

The return of savagery was almost immediate. There was widespread looting and rioting; the old tu-chen's troops, routed, abandoned, left to themselves, turned into murderous gangs.

They robbed flour mills, food stores, and factories. They stole for the sake of stealing. One missionary saw cavalry men with stacks of hats on their heads; they had obviously raided the local hat-maker's shop. The soldiers went on a wild rampage of destruction....

...the gunfire outside the walls of the compound, made us take shelter in the large central hall of the house. The windows by the staircase were barricaded with mattresses. We slept in the hall, and I could hear the sound of shooting that seemed to grow louder at night as if there were a pitched battle going on. The servants were frightened and huddled in their quarters, and seemed to be moaning.

Miss Hodgkin [the tutor] kept asking my father in a quavering voice for assurances that we would not be killed. I can only remember being excited: this was like an adventure story that I had read about being besieged by the natives in Africa....The next morning – and it seemed such a beautiful day after the wild night of rioting – I picked up rifle shells in the flower beds beside the high brick wall.

We were allowed out of the compound to watch the long lines of General Feng's soldiers marching up through the wheat fields toward the city's main South Gate. They seemed better disciplined, better dressed in their grey uniforms than the ragtag troops of the old tu-chen. My mother was impressed: "They even had handkerchiefs," she recalled.

The day was bright and sunny, and everyone seemed relieved and pleased at the sight of these troops, especially the missionaries from the Southern Baptist compound across the mud road from us, who were also out to see them arrive. I waved to Miriam and Hendon Jr.

Their father had gone to the South Gate and would be one of the first to greet the Christian general when he made his triumphal entry into Kaifeng (May 14, 1922). The missionaries of all Protestant denominations, Anglican, Methodist, Presbyterian, and Baptist, hailed Feng as "an example of the power of the Christian message," but the Catholics stood aloof; he was not one of them....

General Feng brought about some much-needed reforms during his time as tu-chen and it must be said that he was a much better governor than

> his predecessor, Chao-Ti. He was a warlord all right, but at that time he was an exception to the run of the other warlords, most of whom had evil reputations as predators and parasites, and overtaxed and tyrannized the people under their control…Feng did attempt to stamp out corruption.
>
> He improved education generally in Honan, he converted many temples into classrooms, and he opened the first public library in Kaifeng. He sought to ban opium smoking, prostitution, and gambling, and tried to enforce the rules against foot-binding, which was still prevalent in the province.[285]

Soon thereafter there was a drought and subsequent famine. The fields in the countryside were parched for months so thousands of desperate country people moved into Kaifeng seeking help. Some even offered their children for sale in order to have money to buy food.

Smallpox and other diseases broke out. General Feng opened camps where people were fed. Some paraded their enormous rain god through the streets so that he would (as was told to Florence) "feel the sun on his head and shoulders and...have pity on the hungry people and give them rain," but nothing happened.

Both Stursberg and Florence gave the following miraculous account in their books. In response to the drought, General Feng had a huge platform built. He had posters mounted announcing a Christian prayer meeting at a specific time a few days later.

Feng invited the missionaries, local pastors, and citizenry to pray with him for rain. Thousands showed up at the appointed time, but the prayer meeting did not last long. During the prayers the skies opened up. Everyone ran for cover as torrents of rain came down.

Feng preaching to thousands of his troops in Kaifeng in 1922 from a high platform.[286]

The Stursberg Compound in Kaifeng

The Stursberg compound, covering several acres, was elaborate for that time and location. The house had electricity from its own generators. It was the only dwelling in the whole province to have flush toilets and both hot and cold running water. Even boasting a tennis court, it must have been a stark contrast to the missionary residences across the dirt road – much less the average Chinese home of that era.

However, apparently the Stursbergs were generous and allowed others – both Westerners and Chinese – to play on that tennis court. In fact, on reading Stursberg's book, it finally cleared up a mystery for me about where in interior China our family photo on that very court could possibly have been taken. (Previously we grandchildren had jokingly called it "Grandfather serving in China" – as if his years of service in China were nothing more than a holiday at a resort.)

Mr. Phil White, Governor Shang Chen, Dr. Hendon Harris, and Dr. Greene Strother on Stursberg's tennis court.

Grandfather Harris serving in China. Reverse of photo reads "H. M. Harris serving in Honan, China. Aug. 1937."

In their lovely home the Stursbergs hosted elaborate parties and programs. Peter Stursberg described one in which the warlord before Feng and his many wives and concubines were entertained.

Stursberg wrote about another party in which he and his brother dressed in goat skins and carried wooden staves for a rendition of "Men of Harlech." However, he stated that he felt outdone that day by Hendon Harris Jr. who dressed as Abraham Lincoln and recited part of the "Gettysburg Address."

Stursberg Compound in Kaifeng. Note tennis court in foreground.

Education of Foreign Children in China

Many of the children of missionaries were sent away at very young ages either to a missionary boarding school or to the home country for schooling. One missionary sent his son home to England for his education when he was only five years old.

In the fall of 1919 Miriam and Helen, then only seven and eight, were sent 300 miles away to a boarding school. After that they returned home to Kaifeng only for Christmas and summer vacations. However, Peter Stursberg, his brother, and Hendon Jr. had private tutors in Kaifeng. There were even music lessons for the children.

Stursberg recalled:

> The Christian general, the medieval setting of Kaifeng, and the province of Honan, and our colonial lifestyle in the Postal Commissioner's House were all part of my education. So were the missionaries and the missionary children like Miriam and Hendon Harris. However, the formal lessons in the children's room, now known as the schoolroom, were a blank in my memory.
>
> What did our English governess, Miss Hodgkin, whom Miriam called "Miss Pigpen" as a girlish joke, teach us? What books did she use? Of course, it was pretty straightforward stuff – reading, writing, and arithmetic. I remember having a den in a cupboard underneath the staircase in the great hall where I wrote and illustrated stories about pirates and American Indians, about whom I knew nothing except what I had absorbed from books, but never a line about the Chinese, about the mah foo or the servants' children with whom we played. I read avidly…books…which had an English setting that seemed ideal although far removed to a boy growing up in Kaifeng.
>
> Still Miss Hodgkin seemed to have done an adequate enough job as she left my brother Richard and me sufficiently well versed to be equal to our peers when we did go to school…. Education was in some ways a heart-rending problem for many foreign families in the Far East.[287]

In Henan the postal commissioner traveled by sedan chair carried on the shoulders of Chinese servants. Other than travel by train, the missionaries more appropriately walked or relied on donkeys or wheelbarrows for transportation.

Medical Maladies and Civil Unrest

There were boyhood accidents including Eugene swallowing an open safety pin, then at another time gashing his cheek when he fell on a broken bottle. Little Hendon was butted against a wall by a cow when he went to see her calf. His head wound had to be stitched shut. Fortunately China Inland Mission's hospital was located in Kaifeng. Most of the time that hospital had at least one doctor.

The Harris family went through bouts of various serious illnesses. These must have weighed heavily on the young Harris couple, especially because their firstborn son had died after a childhood disease. However, there was also the continuing threat of Chinese civil unrest. Father wrote of a time when the family fled on a train and he – as a young boy – was hidden in a Chinese mailbag.

In spring 1925, when Hendon Jr. turned nine years old, he and his younger brothers Eugene and Lawrence all were quite ill with scarlatina when another major anti-foreign movement swept all of China. Both threats – disease and war – hit at the same time. On that occasion most foreigners fled to protective coastal cities, but the Harris family had to wait until the boys were well enough to travel.

However, on arrival in Shanghai, three year old Richard came down with scarlet fever. With his contagious illness no hospital would treat him for fear of infecting the other patients. One doctor said "I'm sorry Mrs. Harris, but there is nothing else I can do for your child." As Richard's fever raged his parents feared he would die. However, after much prayer, persistence, and help from another doctor, he survived.[288]

Second Trip for Harris Family Back to America

In 1926, at the time of the Harris family's second furlough to America, ten year old Hendon Jr. could remember neither his grandparents nor America. China was his home.

Travel to the U.S. from China is equidistant whether one goes east or west. Therefore, that time the family traveled back to the U.S. through Europe. They booked passage on a tramp steamer to cut expenses. Their ship stopped at many interesting ports, took them past India, up through the Suez Canal, through the Mediterranean, and on to Europe. However, food onboard was poor and the children developed sores on their bodies. Florence assumed the skin problem was from malnutrition, so at every port she sought fresh fruit for the family.

Florence wrote about a side trip in one Asian port:

> Once when we were riding near a jungle a large tiger ran across our path chasing a monkey into the underbrush. It was an exciting and frightening experience. The monkey dropped her tiny baby, which one of the rickshaw pullers picked up and placed in his coat pocket.[289]

Florence further noted:

> The Suez Canal cuts through a sandy portion of the desert, and Arabs mounted on camels seemed to be almost in reach of the ship's deck.... Hendon called us all to the deck and pointing to the hazy distance tried to impress on the children that what they saw was Mount Sinai, where God met Moses with the tablets of stone.
>
> Sailing into the blue water of the Mediterranean was gratifying for soon this ocean voyage would end at Genoa, Italy. We thrilled at the sight of the Stromboli volcano erupting at night, the fiery lava streaming down its mountainside.[290]

From Genoa the family traveled to Milan and viewed Da Vinci's famous "Last Supper." From Milan they took a train to Switzerland where they saw the Alps and Lake Geneva.

The streets of Paris were magically full of merry-go-rounds when the Harris family arrived there on Bastille Day. The favorable exchange rate allowed them to buy attractive new clothes at reasonable rates. From there they traveled on to beautiful sights in London.

The excursion continued by ship to New York City where they were greeted by sights of Lady Liberty. Next came a tour of Washington, D.C., before they traveled to their grandparents' home in Birmingham, Alabama. Somewhere along this journey young Hendon's love for travel was born.

The transition back to American life seemingly went well. The family settled again in Clinton, Mississippi, but for a year Hendon Sr. spent most of his time in Louisville, Kentucky, where he completed his doctoral work at Southern Baptist Seminary.

Hendon's doctoral dissertation, "Indigenous Churches in China," marks him as progressive for that era. He believed that well trained Chinese were perfectly capable of running their own churches.

Before the family was able to return to China at the end of their furlough, the Great Depression struck. At that time missionaries who were in the States were asked to resign and take other jobs.[291] Therefore, the Harris family had to remain in America for several more years while Hendon Sr. taught in a college. The work they had started in China continued under Chinese leadership with the help of other missionaries.

Cows Travel to China

Over the years several other Southern Baptist missionaries joined the group in Kaifeng – including more teachers, an agrarian advisor, and one man skilled in animal husbandry. While the Southern Baptist missionaries had various responsibilities, the goal of them all was to share the Good News of Jesus Christ.

In the early 20th century most Chinese did not drink milk. Up until then cattle in that area were for plowing or pulling carts but never for milking. When the cattle were old, they were slaughtered for beef. Through the Sallee's contacts in America and with some of Eugene Sallee's own funds, choice dairy cows were sent to Kaifeng.

The American dairy cows, soon bred with Chinese bovine, provided a valuable commodity, a new economic venture for the Chinese, plus an opportunity for students from poor families to work off tuition by tending the cows.[292]

Hendon Harris Sr.'s Outstations

Annie Sallee wrote about the fifteen rural mission outstations that Hendon Harris Sr. had established by the time of his furlough in 1926:

> Mr. Sallee was asked by the mission to take charge of the fifteen country stations which Mr. Harris had opened and superintended before his return to the United States....Mr. Sallee gladly gave himself to the pastoral duties and aided the rural evangelists as he went from church to church.

> There were discouraging places, but one of the most encouraging churches was at Lien Cheng where there were one hundred and sixty-one members. The Chinese pastor delighted in showing Mr. Sallee his field of labor. The members came from forty-one different villages [hamlets]...In one village the evangelist would say there was only one church member, in another ten or fifteen, in another six,...but the gospel was becoming rooted in the country, and Mr. Sallee rejoiced in this, for he knew it was from such work that one may expect to see China evangelized.
>
> The quarterly meeting in December, 1929, was held in a city about thirty miles from Kaifeng. A heavy snow had fallen, but the ground was frozen and there was little difficulty in reaching the place....
>
> Another snow-storm began, and Mr. Sallee decided he had better return home. It required ten hours to travel five miles. The car [perhaps Hendon's Model-T Ford?] had all sorts of trouble. Mr. Sallee was pushing, pulling and guiding the car... doing all in his power to keep from being stranded on a lonely road at night.[293]

The Harris Family Returns to China Again

Florence wrote:

> In the early months of 1935, some of the missionaries and Chinese leaders in the Kaifeng area wrote Hendon [Sr.], urging him to return to China as soon as he could, for he was sorely needed. One letter, I recall, was from one Chinese pastor who, paraphrasing Psalm 42 wrote: "As the hart panteth after the water, so we pant after thee." This stirred our hearts deeply, for we loved China and were true missionaries in our thinking.[294]

Florence wrote:

> Dr. Charles Maddry, the [new] Executive Secretary, who did not know us replied to them [the group in China asking for the return of the Harrises] in a humorous mood. He surmised that the Harrises were so old they would likely die on their way over, and besides, they had so many children, the Board would have to charter an entire ship to accommodate them. Thus he dismissed the case.[295]

However, between repeated requests from China and an accidental meeting with the Harrises, Dr. Maddry finally agreed for them to return to Kaifeng.

The Harris Family about 1935. First row Florence, Cita, Helen, Miriam, Richard. Back row Hendon Jr., Eugene, Hendon Sr., Lawrence.

When his parents departed for China in fall 1935, Hendon Jr. was in college and therefore stayed in America as did his siblings Eugene, Helen, and Miriam. However, Hendon Jr.'s heart was still at home in China.

Annie Jenkins Sallee captures the emotions of such separations:

> It is not always easy for one to give himself to serve the Lord when he feels called to a foreign land to preach or teach the gospel. Seldom does one find a missionary who feels he has made any sacrifice for the Lord, for though he sometimes walks a wearisome way, he feels that the toils have been as nothing compared to the joy of service.
>
> But no missionary finds it easy to give up his or her children…for four years of college, and maybe a lifetime….The suffering of the parents is poignant, but it is only exceeded by the loneliness of the children in a land foreign to them [their native country] and often unsympathetic.[296]

After being away from China for 9 years Hendon Sr., Florence, Lawrence, Richard, and Cita Harris were welcomed back by Chinese and American friends who waited for their arrival at the railroad station in Kaifeng. One lady, so eager to see them, went to the train station three times before they arrived.

Fellow missionary Eugene Sallee had died in 1931 while back in America fundraising for the mission.[297] Sallee's absence from Kaifeng put further responsibilities on the shoulders of Hendon, who upon arriving was amazed at the deterioration that had taken place.

Students at the boys' boarding school had mutinied and furniture had been carried off–including the chapel piano which was now in the possession of the concubine of a local government official. With forcefulness and clever maneuvering, Hendon Sr. was able to get the piano and other items back.

Florence wrote:

> Dr. Arthur Gillespie, one of the newer missionaries, was in charge of the Bible School. A strong force of…ladies…were holding the lines in their teaching and evangelistic work. Mr. Wesley Lawton [adult son of the Lawtons in Zhengzhou] was helping in the outstation activities. Hendon… with the help of Dr. Peter H. Lee… put into operation a new boys' boarding school…[298]

A chapter of the International Rotary Club was started in Kaifeng, and Governor Shang Chen was president. Hendon became a member and soon became good friends with the governor, other government officials, and prominent people in the community. Later, because of that friendship several times the governor, who eventually became a Christian, intervened favorably on the part of the mission work.

Revival started in the city. Mr. Yin, treasurer of the province, opened his home for Christian services for high government officials. Soon on Sundays the Post Office, several businesses, and some banks closed during worship services. Dr. Peter Lee led the church in fasting and collecting offerings for foreign missions during the World Day of Prayer. Famed evangelist Dr. George Truett held a revival at Kulo Church in Kaifeng. Individuals in the church, including Grace Wu, offered to take more responsibilities.

In 1936 Florence had to have surgery for a fibroid tumor. In early 1937 Hendon had just been offered a high position in Shanghai with the Baptists when he had to have an emergency appendectomy and almost died. Florence wrote:

> Word…spread all over the Christian community. People of every denomination were praying for his recovery. Many of these not only prayed that he would get well, but that he would be kept in Kaifeng...
>
> God heard the people's prayers, for Hendon recovered. He felt led to decline the distinctive offer in Shanghai. God surely was in this decision for only He knew that a terrific Sino-Japanese war was imminent. At one point Hendon was the only senior male missionary left to bear the tremendous responsibility of our [Kaifeng] station.[299]

The Harrises with some of their Chinese students.

Florence recalled that in June 1937 the Harris family was on vacation with several other missionary families in the cooler mountain area of Henan Province when news came of Japanese movement into the area. The male missionaries returned to Kaifeng, while, for the next ten months, many of the women stayed in the mountains near the children's boarding school.

During the winter the boarding school was evacuated and sons Lawrence and Richard Harris left with classmates for Hong Kong. Hendon Sr. returned to see the family only at Christmas and then to see Florence and Cita in April 1938.

Florence wrote:

> While he was on that short visit with us [in April 1938], he received a telegram from the Chinese Christians in Kaifeng, "Dr. Harris, please hurry back. We want you to be with us when Kaifeng falls to the Japanese…Kulo Church Members."[300]

Hendon Sr. instructed Florence to escape with Cita to Hong Kong but said that he had to immediately return to Kaifeng.

Japan, not yet at war with the US, was instructed not to bomb the evacuation trains identified with forty foot US flags on top. These trains had been prepared by US consuls in Hankou to evacuate several hundred Americans. After making their way to Hankou, Florence and Cita left on one of those trains.

When Hendon reached Kaifeng that spring of 1938, most of the wealthier and more influential Chinese had already fled. Japanese planes had dropped bombs on large parts of the city. Hendon Sr. had large American flags painted on the roofs of the Baptist buildings, hoping that would spare them.[301]

Japanese Occupation of Kaifeng

The Kaifeng International Relief Committee, of which Hendon was executive secretary, was composed of men and women, Chinese and foreigners. They knew what had happened in other parts of China and wanted to be ready should the Japanese invade Kaifeng. They had prepared for such a day by setting up a plan and stockpiling tons of wheat.

Florence wrote:

> Hendon asked for and was granted large sums of money with which to buy provisions, fuel, and the necessities for the anticipated thousands of refugees that would flee to our several church centers for protection.[302]

Plans were set in place so that if the Japanese took the city, all the missionaries in Kaifeng representing the various Protestant denominations along with the Catholics would work together to shelter between 14 and 15 thousand Chinese.

When the Japanese entered Kaifeng, Hendon and Brother Claugherty of the Catholic mission, as representatives of the Kaifeng International Relief Organization, were waiting at the city gates. Brother Claugherty assisted Hendon as daily they monitored the well-being of the various refugee shelters.

Hendon is second from right on front row.

The Japanese soldiers exercised their cruelties on men, women, and children and looted or destroyed shops. In some parts of China the Japanese soldiers were promised three days of unbridled looting and rape in reward for taking a city.[303]

The Japanese rules and regulations in Kaifeng constantly changed and those who disobeyed were severely punished.

China Inland Mission reported that the Japanese on their arrival to Kaifeng had ordered the Chinese to "leave their doors open night and day."[304] That allowed the Japanese soldiers access to any and all females and anything else they wanted.

Since their countries were not yet at war with Japan, the confines of the missionaries' compounds were supposed to be safe zones. In theory the flag of each foreign country protected its compound.

Some missions sand bagged their gates. All fought to keep the Japanese soldiers out. It is said that at one point a Japanese soldier entered the Baptist compound where Hendon Sr. was. Women and girls screamed and ran. At six feet two inches, and much larger than the Japanese soldier, the story goes that Hendon grabbed the intruder by the scruff of the neck and seat of his pants and threw him over the compound wall into the street outside – if true, that was a gutsy move.[305]

A bomb fell during one of Hendon's daily inspection trips to the various shelters in Kaifeng. The explosion knocked down a wall crushing and killing a couple crouched nearby. Hendon was only spared because he had rolled into a ditch.[306]

Henan: Fire and Blood states:

> The 1930's and '40s saw the Protestant church in Kaifeng come of age. During the Japanese invasion of 1938, it was stretched to the limit taking care of the 14,000 refugees who flooded into the city from the countryside.[307]

Through Fire: The Story of 1938, China Inland Mission's report, described CIM's effort in Kaifeng:

> Here in Kaifeng plans were made early for each Mission to receive people, and thus when the city fell some 14,000 refugees suddenly rushed into gateways that they hoped would be able to protect them because of the foreign flags that floated above.[308]

CIM reported that most in their camp of 900 were women and children including only 26 church members. However, during those months of internment many placed their faith in Christ. Among those in the Southern Baptist (Kulo Street) Camp were female Chinese evangelists who sparked a revival.

The Southern Baptists' 1939 *Annual* stated:

> ...over fifteen thousand refugees crowded into ten Christian centers in Kaifeng. Baptists carried on three camps, one in the city where about four hundred were cared for and two at the South Suburb Compound caring for about three thousand.[309]

The camps were well organized. Everyone was assigned a job of cooking, cleaning, grinding grain, or teaching a class. At night the church and school buildings were used as dormitories. During the day they served as chapel and classrooms. This went on for an extended period of time.

Florence wrote:

> The 14,000 Chinese in the mission havens of refuge went unscathed. A man and a little girl who ventured out were ruthlessly killed.

> Houses were broken into and their occupants ... killed and the women humiliated.[310]

The missionaries tried to continue their work as much as possible. 1939 *Annual* (report for 1938) stated:

> The Bible School reopened in September with an enrolment of twenty-six. Dr. H. M. Harris, in addition to many other duties, has been acting principal of this school. This school for the training of evangelists of limited preparation is running smoothly.[311]

Meanwhile in Hong Kong, Florence, Cita, and Lawrence all had to be hospitalized for various ailments.

That fall when the Harris children were well and in boarding schools and Florence was strong enough to travel, she decided that she had to return to Kaifeng to see her husband and to do what she could to help.

Florence's story of that trip back to Kaifeng is dramatic and yet full of humor. In my opinion it is the highlight of her book *How Beautiful the Feet*. Her account of that trip is also told in a letter dated November 5, 1938 in the archives of the Southern Baptists.

At that time there were three main fighting factions in China. A Chinese civil war was going on between the Communists and the Nationalists. Both of them were also fighting the Japanese, who invaded further and further into China.

Florence's return to Kaifeng involved de-boarding trains twice to hide from strafing by Japanese planes and a perilous crossing of the Yellow River. Her train was the last to be able to get through to Zhengzhou before the tracks were bombed.

She wrote in a letter at that time:

> The destruction of trains, bridges, railroads and whole towns along the way was terrible....
>
> The train on which I rode was the last one ... the following train was hit by bombs and a few days later the line was entirely out...for half a day my train hid in the hills as bombers were attacking a town close by.[312]

Earlier that year a dam had been breached in an attempt to slow the Japanese advance. The ensuing flood changed the course of the river and resulted in the deaths of an estimated 850,000 to 4 million Chinese.[313]

Zhengzhou, where Florence had arrived by train, was at that time in Nationalist Chinese territory and Kaifeng, only 50 miles away, was Japanese occupied. Train tracks and roads in between had been destroyed. The only way to get back to Kaifeng from Zhengzhou was through no man's land across the treacherous Yellow River that then separated the two cities.

According to *Henan*: *Fire and Blood:*

> On 11 May 1938, General Chiang Kai-shek ordered his Nationalist soldiers to blow up the dyke that contained the Yellow River...just outside Zhengzhou, in an attempt to halt advancing Japanese troops....this brutal measure did not consider the local Chinese population, 900,000 of whom perished in the ensuing flood. For eight years the river flowed unimpeded across the eastern Henan plain. Approximately 3,500 villages and towns were wiped out, leaving some 11 million people homeless and facing starvation. The devastated area was ten to twenty miles wide...[314]

The breach of the dam dumped a huge amount of water onto the plain. According to numerous sources, the river then became several miles wide. Part of the new river had rushing currents; some of it was swampy with an occasional "island" slightly above water level. Boats had to be manually poled across that distance.

Without knowledge of the flow of the constantly changing river and with frequent unexpected debris, like uprooted trees that suddenly popped up, the crossing was long, dangerous, and difficult.

Despite repeated obstacles en route, Florence was able not only to return home to Kaifeng but also to smuggle across Japanese lines $5000 in paper currency for the CIM hospital in Kaifeng. She arrived caked in mud from hiding in ditches. Blue dye from a new hat she had just purchased in Hong Kong for the reunion with her husband ran down her face. Although several other missionaries came in that way, later a Kaifeng evangelist crossing the river was killed.[315]

Under Japanese rule the foreign missionaries were usually allowed to travel between compounds, but rules constantly changed about what the missionaries and Chinese had to do. Most missionaries curtailed their trips outside the city.

One missionary biography reported:

> The grisly stories that came out of "the rape of Nanking" in 1937-38 were repeated again and again in other cities, towns, and villages as the war progressed.[316]

The Japanese chief of police frequently visited the Harris household at night to check on what they had heard on their radio. He described gruesome accounts and said he could not sleep at night unless that day he had killed at least one Chinese.

The Japanese police headquarters was only half a block from the Harris home. The Chinese claimed that no one taken into that police headquarters ever came out alive. Black smoke constantly came out of the chimney and it was rumored that bodies were being cremated there.

When the Americans' mail was not delivered for several months, Hendon Sr. went to the police headquarters to try to find out why. A Chinese postman had told Hendon that the Japanese were trying to read the mail and when they could not, they burned it.

Just for going there to inquire about the mail Hendon Sr. was detained, interrogated for two hours, and threatened with handcuffs. Then he was forced to give a written apology for asking about the mail. To his chagrin, the Japanese proceeded to torture several postmen as they tried to discover who would have told that mail was burned.

Whenever possible the missionaries interceded (several times successfully) on behalf of the Chinese. Often innocent infractions of the Japanese regulations were due to the fact that peasants from the countryside might not know the latest rule of the Japanese.

The face of one elderly woman from the countryside was repeatedly slapped because she did not know that she was supposed to bow to the Japanese sentry – until a missionary intervened and smoothed out the situation. In another incident, a man from the countryside was beaten to death with a golf club by a Japanese soldier.

Dr. R. H. Glover of CIM wrote in *Through Fire: The Story of 1938:*

> The large majority of missionaries in the affected areas…resolutely decided to remain despite the serious risks and sacrifice involved, so that they might minister to the material and spiritual needs of the Chinese people in this their day of terrible suffering, sorrow and loss.
>
> They threw themselves unreservedly into the task of caring for wounded, sheltering and feeding the destitute, comforting the dying and bereaved, and making such service a means of giving the message of God's love and saving grace to those people in their dire affliction and extremity.
>
> Not only did such unselfish and sacrificial service by the missionaries and their Chinese fellow workers bear the immediate fruit of salvaging thousands of lives from death and relieving a vast amount of suffering, it resulted also in leading very many souls to the Saviour, besides bringing rich spiritual blessing to the Chinese Christians, whose faith was strengthened and whose love and zeal were stimulated by their leaders' example and by the mighty work of God which they beheld.[317]

Cita Harris Strunk reported that in addition to her father's many other duties that he at times bandaged the wounded and helped in feeding the masses.[318]

In 1940[319] with only two days' notice all the British and Canadian residents of Kaifeng were expelled. Florence wrote:

> Dr. Walker, an Englishman, head of the China Inland Mission Hospital… [came] to ask Hendon to take charge of the hospital. Hendon could not refuse. When the local Japanese regime attempted to take charge of the institution after the British medical staff were evicted, my husband let

> them know they had better not try to take over an internationally owned hospital...Fortunately, the trained Christian Chinese personnel were able to keep the clinic and wards in full operation until near the Pearl Harbor incident, when Hendon left China.
>
> When the Japanese took over Kaifeng, the Y.M.C.A. head, who was an American... asked Hendon to accept oversight of [his] ...building. Here again Hendon was able to keep the Japanese [from taking possession]. But the management of the two additional places was a tremendous responsibility.[320]

The Chinese were capable of running those institutions. However, by transferring directorship at that time to Americans, the Japanese were kept out.

Later when the Americans left Kaifeng, a Swiss missionary was able to oversee the CIM hospital and thus kept it out of Japanese hands for rest of WWII.[321] (The Swiss never entered WWII so Japanese had no grounds to take their property.)

Soon after the British left Kaifeng, the Japanese conspired to expel the Americans from that city. A Chinese Christian overheard the plot and notified Hendon. Florence wrote:

> That same night, though it was dark and late, Hendon lost no time in contacting [missionary] Miss Thelma Williams, whom he pressured into leaving at once for Chengchow, where she was to telegraph the American consulate.[322]

International attention from around the world was immediately on Kaifeng. The very next night the Harris family heard radio broadcasts from London and Manilla referencing the plot.

Stirred up high officials from America and Japan bombarded Kaifeng with questions. The local Japanese officials were stopped in their tracks. Quick action saved the day. Until the bombing of Pearl Harbor, when Japan was officially at war with America, Americans who chose to do so were allowed to stay in place at their posts in China.

That account is repeated *In War and Famine: Missionaries in China's Honan Province in the 1940's* which relates that in 1940 the British were ordered out of Kaifeng and that in 1941 Americans in Kaifeng were also supposed to have exited, but mysteriously the exit order for Americans was rescinded.[323]

Another missionary biography quotes a letter by Florence Harris to the Foreign Missions Board, published in the Southern Baptist *Annual* 1940:

> The middle class and poorer people have gradually drifted back to their homes in occupied areas, but on the whole, the student class, the intellectuals, and the wealthy ones who fled have not returned. Cities and hamlets in

> unoccupied areas continue to be bombed by invaders, thus causing much destruction to Baptist property.
>
> The lives of our missionaries have been miraculously spared during these perilous times. Traveling on railroads is made exceedingly dangerous by Chinese guerillas who also make raids on city suburbs and depots, causing anxiety and destruction. Bands of desperadoes, taking advantage of the present disturbed condition, swoop down on the poor farmer folk, taking their little. Thus the winter is faced with untold suffering among this people.[324]

The *Annual* 1940 discussed relief work:

> Dr. H. M. Harris, as Executive Secretary of International Relief Committee, has dispersed about two hundred thousand dollars to the tens of thousands of flood and war victims around Kaifeng's fourteen counties.[325]

The *Annual* 1941 reported:

> Dr. H. M. Harris, executive secretary of the Kaifeng International Relief Committee, reports the expenditure, during the last three years, of well over half a million dollars, Chinese currency, for the relief of distress caused by the present conflict.[326]

Hendon Sr.'s relief ministry in China was so outstanding that over the years the Chinese Government decorated him three times.[327]

Remarkably, during that period several new missionaries and others who had been on furlough outside of China came back in – traveling with their families into or through Japanese occupied territory.

Dr. Arthur Gillespie and his family returned to Kaifeng in 1939. Both Gillespie and his wife had taught at the Southern Baptist Kaifeng station for six years starting in the early 1930's while their young twin sons had been attended to during the day by a Chinese amah. Following that term in China the Gillespies extended their one year furlough in the US to two in order to allow Arthur time to finish his doctorate.[328]

By 1939 when the Gillespies returned to China they had three sons. Their son James later wrote:

> The American Consulate official urged us not to travel to inner China under the Japanese, and would only grant travel permits once our parents signed statements that they would not expect the United States to take action in our behalf should we encounter trouble.[329]

Chinese caption reads, “Group photo was taken on April 25, 1939 at the spring baptism for newly baptized church members and their pastors of the Baptist Church of the Drum Tower Street of Kaifeng City.” On the back of the photo is written “Baptisms by Peter Li [Lee], 102 - Li, HMH, Mrs. HMH at right side.”

By then the Baptists owned multiple compounds in or near Kaifeng. There was the large Baptist compound within Kaifeng where a church and schools were located. Outside of the city, where the Gillespies lived, were two more Baptist compounds separated by a dirt road. The Lawtons with their young daughter lived in the same compound with the Gillespies. Across a dirt road was another compound in which the B. L. Nichols family lived with their two sons.

Baptist missionary families in Japanese occupied Kaifeng in Fall 1939. Front row: Joyce Strother, Buford Nichols Jr., Joe Willis Strother, Norman White, Paul Gillespie or John Nichols?, James Gillespie, Arthur Gillespie Jr., Attie Bostick (adult), Paul Gillespie or John Nichols?, Wesley Lawton (sitting) Second row: Martha Strother, Phil White, Mattie White, Jeanie Jo White, Mary Frances Nichols, Maudie Fielder, Muriel Lawton, Zemma Hare. Back row: Greene Strother, Florence Harris, Hendon Harris Sr., Arthur Gillespie Sr., Miss Clifford Barratt?, Pauline Gillespie, Josephine Ward, Wilson Fielder, Buford Nichols Sr., unknown, Grace Stribling, Annie Jenkins Sallee.[330]

Three of the missionaries in the photo above were dead within two years and prior to the bombing of Pearl Harbor. Muriel Lawton died in 1939 and Zemma Hare and Phil White in 1941. All succumbed to various illnesses. The stress at that time must have been tremendous. One man, who was one of the young boys in the photo above, wrote that the rest of his life he and his father both suffered post traumatic stress whenever they saw military personnel.

China Baptist Theological Seminary in Kaifeng

For the first two years of Japanese occupation of Kaifeng, Hendon Harris Sr. had been the only male Southern Baptist missionary there.[331] The Gillespies returned to China 1939 and the Nichols family came to Kaifeng that same year with a larger mission in mind – to help Dr. Harris in starting the China Baptist Theological Seminary.

Florence wrote:

> As far back as the 1920's Hendon dreamed of such a school. He talked to individuals and even invited interested Baptists from other sections of China…to visit us in Kaifeng to discuss this burning desire for an all China seminary.
>
> Each of the four China Missions had its own seminary, but Hendon visualized a central, outstanding institution for higher learning, comparable to the Southern Baptist Seminary in Louisville, Kentucky.[332]

The 1940 *Annual* stated:

> For more than ten years they have prayed and planned for a seminary which could meet the need for advanced theological students.
>
> In the spring of 1938 a group of Chinese Baptists and missionaries met in Kaifeng…Three days were spent in conference and prayer….after much prayer and supplication, it was unanimously agreed, "We have been so impressed with the manifest leadership of the Holy Spirit in these meetings that we feel we cannot longer delay the beginning of the seminary in Kaifeng."[333]

Florence wrote:

> Kaifeng…was under the Japanese, and the fighting between them and the Chinese was still in progress. In fact, at the very time that… the Chinese and missionary representatives from all over China met for conference and prayer concerning the proposed seminary, a serious skirmish between the two warring factions threw the city into great turmoil and fear.[334]

Florence then quoted a letter written by Hendon:

> "We cannot wait for things to settle down. From the Christian standpoint things have never absolutely settled down. Not in Nero's day and maybe not in our own day. Great institutions are born in adversity. Let me quickly add that our seminary does not have to be large in number to be a great school."[335]

Florence indicated:

> Dr. B. L. Nichols stated in the May, 1940, Commission: "A temporary constitution was drawn and adopted. Dr. H. M. Harris was selected as chairman of the faculty...Approval was expressed of those already designated ...as members of the faculty – H. M. Harris, A. S. Gillespie, Peter H. Lee, and B. L. Nichols...and Miss Ola Lee [sic] was nominated by...[Women's Missionary Union] as Dean of Women."[336]

The courses included all the standard master level seminary subjects including Greek and Hebrew. James Gillespie stated that his father taught Greek and theology and wrote:

> Because the seminary was to be fully accredited the Chinese had to learn to translate the Bible from the original Greek and Hebrew. Because the Chinese language does not conjugate their verbs, the students had to know English in order to study Greek....Father and the students would go out into the villages and preach.[337]

Faculty on first row: Miss Ola Lea, Dr. Gillespie, Dr. Harris, Dr. Nichols, Dr. Peter Lee. Students are on second row.

Hendon Sr. wrote first to field director Dr. Rankin in January 1940 and then to executive director Dr. Maddry in March about his son Hendon Jr. who was slated to finish seminary at Louisville that spring. On January 1, 1940 Hendon Sr. wrote to Rankin:

> Our son Hendon Jr. writes that he has sent in his application for appointment. He gets his Th.M. this next May. He is interested in evangelistic work and we need a city evangelistic worker in Kaifeng badly. Members of the seminary faculty can at most give passing attention to the work of the churches. So I hope he can come here or at least to Interior China.[338]

Maddry replied on April 22, 1940:

> I had an interview with Hendon Jr., when I was in Louisville some several weeks ago and we talked over the matter of his coming out to China as a missionary. One trouble with reference to his appointment is the fact that he is still rather young [He had just turned 24.] and, then, too, he does not have selected someone to come out with him as a helpmate. [He was unmarried.]
>
> If he were to come to China, it would not be possible for us to send him to Kaifeng because it is a policy of the Board never to put members of the same family together if we can help it.[339]

The Kaifeng seminary opened its first year relatively peacefully with six students, who came from north, south, and central China. The second year 18 enrolled. It was during this second year that trouble brewed with the Japanese. Florence stated:

> ...local authorities ordered everyone to be vaccinated by their own (Japanese) prescribed medical staff in a certain building. It was of little concern to the Japanese...if those in charge of the vaccination chose to use the same dirty piece of cotton to swab six or ten arms....
>
> When one of the seminary students dared to question the use of a filthy piece of cotton on his arm, he was speedily singled out and roughed up... This Mr. C. K. Chang was taken off and put in confinement for several weeks.
>
> During this imprisonment, the young theologian was questioned as to his education and attainments. Then the Japanese offered him a high position with accompanying attractive salary if he would leave these foreigners and come to work for them. Mr. Chang's answer was that as a Christian he felt compelled to follow God's call to the ministry.
>
> "So you are a Christian, eh?" taunted the officer. With that C. K. Chang was forced to grasp a red-hot poker that had been prepared for him.....[Later] When Hendon saw the scarred hand of his student, he asked to be allowed

to put his finger on the wound. He said, "I want to touch a scar that was made because of your faithfulness to Christ."

At the close of the seminary session in the spring of 1941 Hendon left for the United States, so the seminary went into other administrative hands.[340]

The account of the seminary student's hand being burned and his subsequently becoming a zealous evangelist was also told in *The Commission,* January 1941.[341]

Missionary Families Ordered out of China

Missionary son James Gillespie wrote:

> On October 12 [1940] the American Consulate sent a message...saying that the State Department was advising all non-essential American men, women, and children to leave China at once.[342]

Missionary families, including the men, soon went to Shanghai preparing for departure, but a debate followed. James Gillespie later wrote:

> Dr. Maddry, Executive Secretary of Southern Baptist Foreign Mission Board, had sent a cablegram indicating that the Southern Baptists in America would not tolerate the evacuation of able-bodied men and other missionaries with no children in China. It became evident that women with children would leave their husbands behind...Father tried to explain to us that business people, specifically those who worked for the Tobacco Company, felt that they were in essential positions in China and would remain there. Father explained that the work of the Lord was much more essential than that of tobacco people....I cried and cried.[343]

Erleen J. Christensen, daughter of Lutheran medical missionaries wrote that her parents with 2½-year-old Erleen in tow, arrived in China in September 1940.[344]

Rejecting the order of the US embassy soon after their arrival for American women and children to leave China,[345] Erleen's parents decided instead to exit language school and immediately go even deeper into China.

They crossed Japanese occupied territory to get beyond Japanese lines in far western Henan Province. Another missionary family started out with them, but at one point took a slightly different route than the Christensens so as not to attract attention by too many traveling together. That other male missionary was shot and killed en route by bandits.[346]

That fall of 1940 just before the edict for American women and children to evacuate China, the Harris family was having their own crisis. Cita, age 13 and the only Harris child still in Asia, was on her way back to a school for foreign children in Japanese occupied Korea. Cita and the students accompanying her were supervised by several teachers on that three day trip which involved several modes of transportation. One leg of the journey was by ship.

Florence wrote:

> Cita and another girl her age were on the deck alone when against their wills, they were forced to enter one of the ship officers' rooms. There were a number of uniformed Japanese men sitting about. The girls were told they were going to be given [a physical] and antitoxin. They protested and tried to explain that they had already received their necessary shots.
>
> Alone and unprotected at the moment, each child was made to disrobe, and while she was lying on a table, an … injection.[Cita explained that they were forced to take off their panties and were given an injection in the anus.]
>
> In a week or two we were startled to receive a cablegram from Cita's physician urging Hendon and me to come at once to Korea, as Cita was very ill.[347]

Repeated cables back and forth did not reveal the diagnosis. Later the doctor explained that if he had written in the cable the true diagnosis (typhus), Cita would have been moved by the Japanese to a pest house where surely she would have died.

Just as Hendon and Florence were preparing to go to their daughter's side, the order came for all American women and children to evacuate China and Korea.

On October 17, 1940 Hendon wrote to Dr. Maddry, asking for money to pay for Florence and Cita's return fare to America. When it did not come in time, the North Presbyterian Board loaned the Harrises $480 for Florence and Cita's passage to the States.[348]

According to Florence, on one leg of Hendon and Florence's trip to Korea to join Cita they were told by the Japanese officials that their travel documents were not correct and they were put off the train in the middle of the night in Manchuria. The Japanese officials then took all their money except for enough for two third class tickets to return to Beijing. Back in Beijing they were able to get the right documents and borrow more money to proceed.

The US government had already set the date for all ships carrying American women and children to leave Asia. When Florence and Hendon first reached Korea, Cita was initially too weak to travel. However, Florence, against the advice of nurses, kept exercising Cita's legs.

By the day they boarded their evacuation ship Cita was able to walk. The Japanese lined up the departing Americans in columns and made them march aboard. Hendon Sr., as with other male escorts, was not allowed to help his family board. Cita was not even permitted to give her father a farewell hug.

The 1942 *Annual* shows that on November 21, 1940 Florence Harris, Mrs. Gillespie, Mrs. Nichols, and numerous other female Baptist missionaries "left the field."[349] They exited from various ports. The names of the children, who accompanied their mothers, are not listed in that report.

On November 18, 1940 (only three days before the ship sailed) Dr. C. E. Maddry, responded to Hendon Sr.'s letter of October 18 in which he had asked for financial help in paying the passage. Dr. Maddry noted that Florence and Cita were already on the passenger list and stated:

> …about 120 mothers and children and missionaries… will be sailing from China shortly. This is going to work a tremendous hardship upon this Board. It is going to cost us close to $50,000 and, frankly, I do not know where the money is going to come from.
>
> ….I sincerely hope that you will stay right there and go on with your work no matter what happens….there is a wide spread feeling that the **missionaries ought to stay there and suffer with the Chinese**. [emphasis added]
>
> ….We are all going through anxious days here now with the war disrupting our work in Europe and Africa as well as in Japan and China.[350]

Apparently in a moment of stress, Dr. Maddry forgot that God was his provider. With God there is never lack of provision. Those funds did come in. (The International Mission Board Timeline shows that December 9, 1943 "Dr Maddry announced that the board was debt free for the first time in more than a generation.")

On December 25, 1940 Hendon responded to Dr. C. E. Maddry's, letter of November 18:

> I can appreciate your anxiety concerning the effect on the Board's receipts of a large number of missionaries evacuating from China. However, a high proportion were children and I have been told that **we still have 134 missionaries in China** [emphasis added].
>
> ….Numbers of these devoted servants of the Lord [still in China] are willing to make any sacrifice, even of life itself, if be the will of God. In this connection I may say that a vast amount of the difficulties [in China] has not been reported at home at all.
>
> ….Last May I was worn out and nervous from the long tension and heavy responsibilities but I am better now. Only the Lord knows what is ahead of us.[351]

Hendon told Dr. Maddry that because of the situation they were still living under, many hardships the missionaries had endured purposely went unreported to America. It does not appear that he even reported what had happened to Cita.

Then Hendon stated that he was writing his letter on Christmas Day. Alone and without any family, he had been invited to go to Christmas dinner, but he could not go out on the street because Kaifeng was under martial law. He wrote: "Inside this city I am sole Protestant missionary of any Board. The others are outside."

In that letter Hendon also warned:

> The time may come when our presence may be as a red flag bringing hardship to our Chinese brethren instead of protection. That time has not yet come. Our work is in a flourishing condition.[352]

The situation with the Japanese continued to decline. By Spring 1941 Hendon was exhausted and wanted out of China. Florence wrote:

> Back in Kaifeng, Hendon tried to persuade Dr. [M. T.] Rankin [Secretary of the Orient]... to get all the missionaries out of China, especially the women, for he was well read on current trends, and had been given a strong hint by the Kaifeng chief-of-police that Japan would inevitably war against America. He told the secretary that he did not feel led to stay and be interned in a concentration camp. Dr. Rankin could not see eye to eye with Hendon.[353]

In a letter dated April 20, 1941 (retained in the archives of the Southern Baptists), Hendon went over the head of Rankin, his immediate supervisor, to address Dr. Maddry again. The two page letter, clearly hiding another message, started out in a friendly, chatty way. "Some 1300 converts were received last year in the Interior China Baptist mission." Hendon reported that he had preached 24 times in various cities during the past two weeks.

Buried further down in the letter to Maddry it indicated that Hendon had just recently had a "lively" face to face discussion with Rankin. Hendon wrote: "Rankin...wants as many missionaries to stay on the field as possible."[354]

Southern Baptists records indicate that on "two occasions [Hendon Harris Sr.] was a recipient of medals for valorous work done in times of stress."[355]

However, in this April 1941 two page letter to Maddry, Hendon Sr. the brave missionary who had been through so much, pled that it was time that he and the other Baptist missionaries be allowed to exit China.

Hendon wrote: "Last May I was in an exceedingly nervous state and remained so for six or seven months." (The Japanese imprisonment and torture of the seminary student, the incidents surrounding Cita's molestation and illness, and Florence's and Cita's evacuation all occurred during that period.)

Hendon had witnessed firsthand the cruelties the Japanese inflicted – especially on those who challenged them. If Hendon were interned, he would surely be a target. His letter went on:

> Here in Kaifeng, we have to deal with an individual who is notoriously savage and abusive. He has been insulting and taken a threatening attitude.

> If I come home, I shall rest for a time. Then I shall be available for conferences, conventions and general deputation work. Just now I do not know where I shall stay in America. I must see Mrs. Harris and we shall then decide.
>
> **...the international situation grows ever more acute...there is a ninety per cent certainty of conflict spreading....It would be a black eye for the Board if missionaries are caught and interned in spite of the warnings.** [emphasis added][356]

Since this letter had to exit across Japanese lines, Hendon could not name in it his intelligence sources concerning planned Japanese aggression against America. However, two weeks earlier, he surely had discussed them with Rankin. It should have also been clear to Maddry that Americans would not be interned by the Japanese unless Japan was at war with America.

(Hendon's warnings of Japan's plans to attack America were spelled out again later in an "I told you so" letter to Maddry dated December 9, 1941 – two days after the bombing of Pearl Harbor).[357]

Finally in Spring 1941 Hendon was given permission to leave China. From the time that his wife and children left Asia in late 1940, Gillespie talked about needing to get back to his children. Certainly their tearful exit would have weighed heavily on him.

According to the 1942 *Annual*, on June 13, 1941 Hendon Harris and all four of the original American faculty members of the Kaifeng seminary left China – all except for Dr. Gillespie.[358] Perhaps it was because Gillespie's scheduled furlough date was furthest away of any of the missionaries, and it was felt by Richmond that someone needed to stay in China to protect the Southern Baptist properties.

Florence wrote:

> Six months after I reached Memphis with Cita, Hendon left Kaifeng. He had been under tremendous strain. In fact, he had faced distressing ordeals for the past six years besides undergoing an operation and an illness that all but took his life. Then there were the daily pressures of the ever-changing Japanese rules and their unpredictable attitudes, while having the responsibility and oversight of the seminary, China Inland [Mission] Hospital, and Y.M.C.A. building as well. He was worn and near exhaustion and needed to return to the United States for a rest period....[359]

It appears that Rankin "abandoned ship" before his "crew." Although he did not want the missionaries to leave, and it was not yet time for his furlough, Rankin left Asia on August 15, 1941 – before the bombing of Pearl Harbor.[360]

Although Maddry's letters insinuated that there were insufficient funds to return missionaries to America, the *Annual* shows that meanwhile other Southern Baptists missionaries kept entering China, some as late at October 17, 1941.[361]

Hendon Sr., Florence, Cita, Lawrence, and Richard in January 1942 in Mississippi.

Pearl Harbor was bombed on December 7, 1941. In less than 24 hours after that "Day of Infamy" American missionaries in Japanese occupied areas of China started becoming prisoners of war – including Dr. Gillespie and numerous others, many of them female.

Christensen wrote that after the bombing of Pearl Harbor:

> In eastern Honan…the internment of missionaries proceeded erratically and unpredictably [for]…just over a year. There were a large number of Protestant fields east of the Yellow River…[including Kaifeng] – CIM, Canadian Anglican, Southern Baptist, Free Methodist…Lutheran…and Mennonite. In…Kaifeng, the [Catholic] bishop and two priests had been killed "by bandits" in late November 1941.[362]

After Pearl Harbor, despite the danger, repeatedly Chinese Christians collected food and money to help the foreign missionary internees. Attie Bostick, serving with the Southern Baptists in Henan wrote: "An evangelist WALKED [emphasis hers] in twenty miles and brought me sixty eggs. His congregation had given money to buy them…"[363] She reported similar kindnesses time and again by various Chinese groups.

Ironically, on December 7, 1941, Dr. Maddry (who had insisted that missionaries stay in China despite the hardships), witnessed firsthand the bombing of Pearl Harbor and the fury of the Japanese. He and his wife had gone to Honolulu to survey missions work there.[364]

In his book of missionary stories titled *Christ's Expendables* Maddry recalled the bombing of Pearl Harbor as: "horrible beyond human imagination…waves of bombers…came over the city [Honolulu] from 7:55 a.m. until 9:15 p.m."[365] That strike caused: "3581 casualties, 188 planes destroyed on the ground 8 battleships sunk or run aground."[366]

Maddry was thankful that he and his wife were not hurt and that only a few days after the bombing of Pearl Harbor they were able to make a speedy exit back to the mainland. The American missionaries in Japanese occupied areas of Asia were not as fortunate.

The January 1942 Southern Baptist periodical *The Commission* showed individual photos of 108 Southern Baptist Missionaries still in war zones in Asia after Pearl Harbor. Of those 98 were in China.[367]

The Southern Baptist International Mission Board Timeline states:

> December 9, 1943, Dr. M. T. Rankin [Secretary of the Orient] reported that forty missionaries from war-ravished China arrived in New York aboard the S. S. Gripsholm.[368]

That "coup" by Rankin in December 1943 of repatriating forty missionaries catapulted him in popularity. Less than six months later he was elected Executive Secretary (head) of the International Missions Board.[369]

The internment of the missionaries could have been avoided in the first place if Rankin had listened to Hendon Harris, the soldier with "boots on the ground" – who with information from a reliable Chinese source had earlier warned of Japan's intention to go to war against America.

Dr. Gillespie was fortunate to be among those who were repatriated in the 1943 prisoner of war exchange.[370] Not all the American missionaries made it out during the war. Some died in those Japanese POW camps including a seminary classmate of Hendon Jr., Rufus Gray, who was tortured to death.

Perhaps Hendon Sr. breathed a sigh of relief that his son, Hendon Jr., had not been appointed to China in 1940 as he had requested. Surely Hendon Jr. would have been interned along with the others.

Although mission agencies in America were surely unaware of it then, it has recently been revealed that this secret prisoner of war exchange by President Franklin Roosevelt in 1943 of those on the Gripsholm included American citizens of Japanese descent who reportedly had been kidnapped off streets in the United States and given no due process. Though some of them had never been to Japan before, they were passed off by President Roosevelt as prisoners of war being repatriated to Japan.[371]

In hindsight Maddry and Rankin were both wrong in their decision to keep missionaries in China when the war was looming. Maddry was certainly inconsiderate in how he worded his letter of Nov. 18, 1940 to my grandfather – suggesting that missionaries should stay in China and suffer. However, Maddry and Rankin were both godly men who made wrong decisions. The Bible teaches us that not only does God forgive us, but that we are to forgive others.

Reunion of Hendon Sr. and Hendon Jr. in the US.

We are instructed to pray the Lord's Prayer: "Forgive us our trespasses as we forgive those who trespass against us." If one does not forgive, it only makes him bitter. Apparently Hendon Harris Sr., Arthur Gillespie, and numerous other missionaries were able to forgive. After the war they returned to China.

Hendon Harris Sr. returned to China in Spring 1946.[372] Florence, leaving all her offspring in America, followed in January 1947.[373] Florence wrote:

> On reaching Shanghai that January of 1947, we did not realize that this was going to be the last and most turbulent and chaotic period for all missionaries in China.
>
> Immediately on my arrival, Hendon fell victim to a virulent...stomach disorder...[and] fell unconscious on the floor. That illness was rather prophetic of the disturbed and violent times that were facing us.[374]

Christensen stated that after WWII was over:

> ...A common thread runs through the reports [of the foreign missionaries] – an admiration of the Chinese Christians, the Chinese church, which had proved its mettle through a time of severe testing. The "indigenous church" could no longer be viewed as a child the foreign missionaries were fostering.
>
> It had grown up during the difficult years on its own, and interest in and commitment to Christianity increased impressively while the missionaries were gone. The China Inland Mission, for instance, reported over 1,800 baptisms during the less than two year period when missionaries were not in residence in [Henan] province.[375]

Back Home in Kaifeng

Hendon and Florence were initially delayed on their return to Kaifeng for several weeks by the Chinese civil war still in progress. The Communists repeatedly dynamited sections of the railroad lines which had to be repaired. However, on reaching Kaifeng the Harrises were able to recover several of their household furnishings that had been taken by the Japanese.

Florence stated:

> A table would be found here and the front piece of the buffet drawer there. From a second-hand shop we actually bought back some ceramics that we had owned before...it was a mystery how...the Chinese knew where to locate some of our household effects...the dining room buffet was retrieved minus its drawers (which the Japanese had used for horse troughs.)[376]

Florence felt that it was important to get her home back into order as quickly as possible. Soon their residence was constantly used again both by Chinese and foreigners for various meetings.

After the Japanese were ousted from China at the end of WWII and before the Baptists could return, several of the Baptist buildings in Kaifeng had been illegally occupied by a Chinese relief organization and United Nations Relief and Rehabilitation Administration (UNRRA). Furthermore, other mission agencies complained that the UNRRA group in Kaifeng was painfully slow in getting any relief started.[377] Both of those organizations defiantly opposed giving back the buildings to the rightful owners who sorely needed them to resume their missionary work. It took months of negotiations and maneuvering, but Hendon Sr. finally won the buildings back.

Hendon Sr. then was able to restart the boys and girls boarding schools once again. To do that required repairs, purchasing furnishings, and organization of many details. Dr. Peter Lee and Rebecca Han were by his side in that endeavor. Florence wrote:

> Hendon had worked...for months at this unnerving task, and the strain of it all. . .undermined his health. He was never robust after this ordeal.
>
> . . . .Dr. Greene Strother took the management of the Bible School, our local seminary, thus relieving Hendon, who continued to teach classes there until his departure to the United States.
>
> The Communists, supplanting the Japanese, were growing stronger, bolder, and more insidious day by day.[378]

That summer Brother Claugherty, the Catholic priest who had helped Hendon with the relief effort, daily brought to the Harris family ice blocks (a rare treat in China) which he had saved during the winter.

In exchange Florence gave the Catholics some of her prized roses. Back in the States, Florence was a master gardener and flower show judge. Years earlier she had wheelbarrow loads of dirt brought in to make a garden behind her Kaifeng home and had succeeded in starting roses. Now the roses were a welcome relief to the mayhem in the city.

Fig trees were not supposed to be able to survive the cold winters in Kaifeng. Against odds and with a lot of tender care Florence had also started a fig tree behind the house. Now it was loaded with fresh figs. Over 30 years after Florence left China for the final time a travelogue for *National Geographic* wrote of unexpectedly seeing fig trees in Kaifeng.[379]

One day, during the end of the Harris's final time in China, Addie Cox rode her bicycle over 40 miles into Kaifeng, accompanied by a Mr. He, to see Dr. Harris. Florence wrote:

> [Mr.] He said that there were forty members in the He clan, who were getting their livelihood from a five-acre tract of land. The Communists condemned this family of forty. . .to banishment, requiring that all cattle, farm implements, and household effects be left behind. . .because they were

> capitalist! Couldn't Pastor Harris intercede for them? This...was repeated in all self-sustained families....Several local "landlords," some owning less than an acre, were beheaded by the communists.[380]

Addie Cox had been an extremely brave missionary. In very trying situations over many years she had been reliant only on God and herself. The *Annual* reported that for the initial three years of Japanese occupation Cox had never once returned to Kaifeng – her home base. She opted instead not to interrupt her work in the countryside. There she also faced bandits and the Japanese. She had stayed in China throughout World War II to be among the people she loved.

That day Addie was asking help from the one person she thought might give it – Harris, the man who had spotlighted her and pled for her missionary support at the 1917 Alabama Baptist Convention, the man who for many years was her co-laborer in evangelism. However, even Hendon Harris Sr. could not give the help that she needed.

As the Chinese civil war between the Communists and Nationalists continued, Kaifeng was once again at the battle front.

Hendon Sr. and Florence's Final Battle in China

Florence wrote:

> The...situation worsened day by day. The Nationalist Army in desperation dug deep trenches just beyond the south wall of the Mission school campus. Next they made openings through this brick enclosure for their own protection against the Red [Communist] army, but exposing our residences and schools to a mass of undisciplined National soldiers ... stationed around us.
>
> These lawless, idle, and frightened soldiers were creating a chaotic situation within our premises, making life almost unbearable. As Dr. Harris would start his seminary class or religious service, some of our own people would frantically...[interrupt] relating the last misdemeanor of the Nationalist troops, that were running rampant on...campus. One time Hendon had to stop some soldiers from tearing down the heavy wooden gates.... When soldiers made a fire on the girls' dormitory wood floor to cook their meal, he felt he had reached the limit of his patience....The army decided to use the campus for their daily drilling exercises, thus disturbing the classes. One by one, missionaries were being killed, and these acts made to look like accidents.[381]

On December 12, 1947 Hendon Sr. and Florence heard the boom of the Communist cannons only about 12 miles away. Fortunately that advance missed Kaifeng. Florence wrote:

> Since Hendon's time was mainly used in policing, not teaching or preaching, we decided to leave China until conditions improved.[382]

In an all day ordeal, a fellow missionary drove from Zhengzhou "over guerilla-infested territory" to pick up the Harrises in order to take them back to Zhengzhou where they were to fly out. They arrived at Zhengzhou after dark when the city gates had closed. Miraculously they were allowed to enter.

The next day Hendon Sr. and Florence were airlifted to Shanghai by the Lutheran[383] World Action plane, which according to another missionary report was "evacuating missionaries of all creeds."[384] Previously most of the foreign missionaries in Kaifeng had left China.[385] Addie Cox was among the few who stayed in China until the final eviction of foreign missionaries in 1951.[386]

In the past several years the Nationalists, including Madame Chiang Kai Shek (official title of the wife of the head of the Nationalists), had publicly supported the missionaries and Christians.[387] Madame Chiang as well as her father were educated in America and professed to be Christians. On the other hand the Communists openly opposed Christians of any nationality.

In January 1948, the month after Hendon and Florence left China, three foreign missionaries (including two women) were killed execution style by the Communists.[388] Two months later (March 1948) CIM reported the killing of five missionaries.[389]

China was no longer safe for any of the missionaries. Their presence endangered Chinese Christians. One Chinese man was punished just for relaying messages for missionaries.[390]

There had been systematic attempts to "arrest or otherwise dispose of Christian leaders."[391] In one instance in 1946 in a Communist occupied area, after a missionary surveyed the condition of his former hospital, 26 of the people he talked to were arrested, and 17 of those 26 were killed.[392]

After reaching the United States both Hendon and Florence spoke in many churches and continued praying for China, their adopted home.

Death of Son Eugene

The Korean War broke out June 25, 1950. In September of 1950 Hendon Sr. and Florence received notice that their son Eugene, who was in the US Army, was missing in Korea. That was followed by notice that Eugene had been killed in action on July 20, 1950 – only days after the start of the war. He left behind wife Lorene, and two young daughters – Lynda and Barbara. Eugene never got to see baby Barbara. Eugene's army chaplain described him as brave, tender-hearted, and gentle in his care of the wounded and dead comrades. The whole Harris family mourned.

In her book Florence recalled tender times from Eugene's childhood. When he was first learning to talk Eugene saw a bird on the windowsill and exclaimed, "Dog!"

"No baby, that is a bird," Florence corrected.

"Bird-dog," insisted the tiny fellow.[393]

Eugene – photo taken in 1944.

Florence continued:

> Years later…[Hendon Jr. and Eugene] were playing around Lake Wilson in Clinton when ten-year-old Hendon got into a dangerous predicament in the water. Eugene, standing on the bank, cried inconsolably, thinking that Hendon might drown. He grew up to be tenderhearted and affectionate.[394]

The Final Years of Hendon Sr. and Florence

According to *Morrison Heights Baptist Church: From the Beginning* (2009), Hendon Sr. and Florence were very active charter members when that church was founded in Clinton, Mississippi in 1958. At that time the daytime gathering of ladies of the Women's Missionary Union of the congregation was named "Florence Harris Circle" in honor of her missionary service.[395]

Before going to China their final time Hendon Sr. and Florence bought acreage near Clinton. After they returned from China, it was developed into a subdivision of attractive homes. This wise investment helped support them in their final years.

During the Harris's tenure in China they, and all of the missionary families there repeatedly had their lives threatened by war and disease. Several people close to them had died including:

Pastor Li, Pastor at Kulo church in Kaifeng from typhus 1913

Maude Goddard, teacher of missionary children at Kaifeng, typhus 1913

Dr. S. H. Carr, of CIM who attended Miss Goddard, typhus 1913

Lawton child 1912

The first Harris son – William Powell Harris 1913

Eugene Sallee 1931 – heart problems

Mrs. Wesley (Muriel) Lawton – complications of an illness 1939

Zemma Hare 1941

Phil White – appendicitis 1941[396]

Kaifeng was just one of several Southern Baptist stations in China. Hendon Sr. had opportunity during his tenure there to rise within the Southern Baptist organization to a "bigger and better" position but felt that God wanted him to stay in Kaifeng. His leadership, especially during the last few years, was sorely needed.

Hendon Sr. died August 23, 1961 and Florence on January 4, 1981. My grandparents, Hendon Sr. and Florence Harris, were but two of many missionaries who worked in Henan Province. However, I feel that God used them to play a small part in the story below.

God's Work Goes On

By some accounts millions of Christians have been murdered in Communist China, but Christianity continues to grow there. Breibart news agency reported in 2014: "Christians Now Outnumber Communists in China."[397]

According to *Henan: Fire and Blood,* (2009) today Henan Province has:

> ...the largest number of Christians and is the centre of the greatest and most sustained revival of Christianity, which has lasted more than 30 years.... Millions of people in Henan have come to faith in Christ....this central province...has become an engine room for the spreading of the gospel of Jesus Christ.[398]

Jesus in Beijing discusses some of the "Uncles" (leaders over multiple churches) in the underground church movement in China:

> **Li Tianen**...born in Henan in 1928...his grandfather was converted by the preaching of Hudson Taylor. Li's mother...[was] a full time evangelist...

> [but] died…when Li was nine years old. His paternal grandparents then brought him up until he entered the Hua Zhong **Baptist Theological Institute in Kaifeng** in Henan Province.[399] [Emphasis added]
>
> **Zhang Ronglian**, perhaps one of today's best-known house church uncles in China,… is one of the uncles who was **trained by Li Tianen** [uncle shown above]. Zhang was born in Henan in 1950…His grandfather was a committed Christian.[400] [Emphasis added]
>
> **Peter Xu**...was born on October 6, 1940 in **Henan Province**…his grandparents prayed when he was born that he might become a dedicated Christian leader.[401] [Emphasis added].

While we do not have direct evidence that Li Tianen and through him Zhong Ronglian are connected to the Southern Baptist seminary and Bible training centers in Kaifeng (both started by Hendon Harris Sr.), there were no other Baptist ministries or seminaries in Kaifeng. Likely some of the Chinese continued seminaries there after the missionaries left.

Hendon Harris Sr. prophetically, ended his 1927 doctoral thesis, "Indigenous Churches in China" with these words:

> The Chinese…[have] many virtues: otherwise they would not have survived to the present…nor would they have been able to make such notable contributions to art, science, literature, philosophy, invention and manufacture as they have….Among the outstanding traits of the people are patience, perseverance, energy, reasonableness, loyalty, reverence for parents, urbanity and sociability. Often are missionaries heard to say: "I like the Chinese and want to be with them." They are very appreciative for kindnesses shown them and already the Chinese churches have produced true saints….Chinese are going to make their contribution to the Christian religion both in institutional and social adaptations and in the expression of the personal experiences of religion and appreciation of the person of Christ.
>
> Perhaps one of the most marked features of Chinese life is…power of endurance; they know how to suffer and bear the yoke; millions understand the bitterness of poverty, death, disease, and inexpressible misery. They may yet show the world a new way to "bear hardness as good soldiers of Jesus Christ." The story of Protestant Chinese churches is not a long one but already many of its pages are written in blood and blotted with tears.
>
> The Chinese have been the civilizers of Eastern Asia; what if they should become the Christianizers of the whole continent? It is not impossible that before another century shall have rolled by Chinese Christianity will have become a major force in the whole world?

When one contemplates the possibilities and the potentialities of the churches that are to be in the Yellow man's land, words appear inadequate to portray what they may mean to the generations yet unborn both in the Far East and the entire planet; the stored up spiritual energies of this most ancient and populous nation will then be released to pour out streams of blessing.

The Missionaries, to the Chinese who live to see such a day, will be humbly grateful for having had just a small part in bringing about such a consummation, even as the present generation of missionaries greet that time from afar: they, too, will see of the travail of their souls and be satisfied.

Surely God has hedged in this great people with mountains, deserts and seas and preserved them to the present time for some glorious purpose. The gospel seed has been sown in hope and the harvest will be sure and bountiful.

Someday the Chinese churches will be sending out their own missionaries in large number; they will see the needs of others and will respond generously. Christ is entrusting his gospel to them as he did to the early disciples and to the Chinese also he speaks as he did of old, "Ye shall be my witnesses." [Acts1:8][402]

The Third Generation: Missionaries to Taiwan and Hong Kong

Hendon Mason Harris Jr. and Marjorie Weaver Harris

In the earlier sections of this book I gave the accounts of the first two generations of missionaries in my family as a background in an attempt to explain my Father, Hendon Harris Jr., a talented but complex man. In addition to being a missionary he was a poet, song writer, screenwriter, lover of history, and in later years a map collector. Because he was born in China both he and the Chinese considered Hendon Jr. to be a "son of the soil" – a brother.

I never met my Powell great grandparents who died before I was born. I barely knew my Harris grandparents who even when we were in the States lived many miles away. Of course, I knew my parents well. Therefore, this third section of the book is written from a different perspective than the previous two.

As stated earlier, Hendon Harris Jr. was born in 1916 in Kaifeng, Henan Province, China, to missionary parents. Like his mother and her father, he was also a great story teller, highly intelligent, and dedicated to his calling. He knew his Powell grandparents, but he never lived close to them. Separation from close relatives is one of the sad facts of life for most families of foreign missionaries. Nevertheless, this poem below expresses Hendon Jr.'s thankfulness for his heritage.

"Ancestry"

Laughter and Tears
By Hendon Harris Jr.

How happy is that man
Who can
His parents scan
And say "I'm not ashamed
From such ones to be named."

Who may their lives review
So true
And own his due
Proud to admit the debt
In heaven, or living yet.

Whose father like a King
Did bring
The royal ring
Of truth to all he did
Alone, or accompanied.

Whose mother's gentle grace
And face
In every place
Shone for her Savior's cause
And kept His laws.

Happy indeed is he
That's free
For all eternity
Because his parents prayed
'Til he was white arrayed.

After returning to the United States from China, Hendon Jr. tested so highly that at age 11 he was enrolled in high school. It was said that his mother sent him to high school in little boy shorts and knee socks while the other boys in his class wore long pants.

Chinese Relationship to Native Americans

Hendon Jr.'s grandfather, William Powell, had been interested in the origin of Aztec Indians. He wrote that he had seen numerous Aztec artifacts in the far flung locations that he visited in Mexico and had even purchased some, which he forwarded to a university. Of course, at that time William would have been unaware of later findings listed in *China: A History* (2009): "recent discoveries [were] made... around Chengdu [China]...[including] a bronze statue... 8.5 feet tall...dated to 1200 BC, ...with features more Aztec than Chinese."[403]

Hendon Harris Sr. had also been interested in the origin of Native Americans. In the middle of his doctoral thesis he stated:

> Anthropologists and archaeologists are raising the question as to whether the Yellow and Red races are really separate races or not: possibly the civilization of the Pueblo Indians, the Aztecs, Mayas and Incas of America may flow from an ancient Chinese fountain head.[404]

Hendon Sr. submitted that thesis in 1927 when Hendon Jr. was 11 years old. In my opinion there is no way that the startling prospect of a relationship of the Chinese to ancient Native Americans would not have been discussed at the Harris family dinner table. Inquisitive 11-year-old Hendon Jr., who was starting high school, would not have missed that interesting tidbit.

Previously I wondered what my father meant when he wrote concerning his interest in Native Americans: "From the time I was a boy in China, until now, I have collected information."[405] With fresh insights stemming from my grandfather's thesis, I no longer wonder. Truly, his interest was sparked very early.

Hendon Jr. was sorely disappointed when, because of an economic downturn in America, the mission board did not have the funds to send his family back "home" to China at the end of their furlough. His parents, Hendon Sr. and Florence, along with other missionaries then in the States, were asked to resign until the economy improved.

Hendon Jr. Runs Away

Hendon Jr. and his sister Helen were slated to spend the summer of 1931 with their grandmother in Birmingham, Alabama. Hendon Jr., age 15, had a job selling Bibles. When people could not pay for the Bibles they had ordered, Hendon Jr. became so discouraged over what he perceived as his own failure that he ran away from his grandmother's home. His family was frantic but could not find him.

Three months later, all the way across the North American continent, a disheveled young man was sitting on a park bench in Los Angeles when an elderly man spoke to him. The man asked the boy where he was born and was shocked to hear the reply, "Kaifeng, China." The man – a former missionary to China – knew Hendon Sr. and Florence well and readily helped reunite parents and their fifteen-year-old son. Hendon Sr.'s only request to Hendon Jr. was that he never discuss with his siblings his adventures during that absence.

It turned out that Hendon Jr. had "ridden the rails" (taken illegal rides on freight trains) those months and experienced some harrowing times. In that economically depressed era many unemployed men roamed the US countryside looking for whatever work they could find. Hendon Jr. told about seeing a man riding on top of a train beheaded by a tunnel when he raised his head. In 2010 Father's brother, Lawrence, confirmed hearing the tale about the beheading and told me that during part of that time Hendon Jr. had been a stowaway on a banana boat and after eating many bananas grew sick of them.[406] Regardless, Hendon Jr. had learned to travel and survive on little, an experience which served him well in later years.

Hendon Jr.'s Education

Aspiring to be an attorney, young Hendon started college at 16 years of age. However, he floundered badly his first year. That summer after a church meeting he placed his faith in Christ. Almost at once he felt a call to the ministry. Everything in his life changed.

His faith in God was personally his own then – not just that of his parents. He met life with new zeal. Hendon Jr. had seen God at work in the lives of his parents, so he now believed that he, too, could trust God.

Understanding Hendon Jr.'s simple faith in God is pivotal to understanding his relationship to the historical maps that he would later find. He explained in *The Asiatic Fathers of America* that his faith in God was reflected in his confidence in the ability of the Chinese people to have made such early trips to America.

Hendon Jr., believed that man was created in the image of God and was confident that early people were intelligent beings. Therefore, he believed that the possibility of Chinese crossing the Pacific to America by 2000 BC would not have been too large a task.

At the time he made his profession of faith, a lady at that meeting promised to pay his tuition for him to attend Columbia Bible College for a year, which he described as a wonderful spiritual experience. However, he earned his bachelor's degree from Hanover College in Indiana where his father was a professor. Hendon Jr.'s three year Master of Divinity degree came in 1940 from Southern Baptist Seminary in Louisville – the same seminary from which both his grandfather and father graduated. He earned his doctorate from Northern Baptist Seminary in 1945.

1940 was also the year that Hendon Jr.'s application to be a missionary to China was turned down by the Southern Baptist Foreign Mission Board, under which both his grandparents and parents had served. He was denied because he was unmarried and considered too young.

It was a crushing blow to Hendon Jr. not to be accepted by the Southern Baptist Board then, especially when one of his seminary class mates (who was married) was accepted for work in China at that time. Hendon Jr.'s frustration over their refusal to take him on then as a missionary is expressed in the tongue in cheek rejection letter to the apostle Paul from the Foreign Mission Board found in his *Famous Unwritten Letters.* That rejection missive to Paul is included at the back of this book.

Marjorie Weaver

Marjorie Weaver in 1940.

Perhaps the Foreign Mission Board's denial of Hendon Jr. in 1940 for being single was why later that same year Hendon Jr. married my mother after knowing her for only three months. However, my mother Marjorie Weaver, also felt called to China.

Marjorie's parents were both Canadians by birth. Born in Rochester, New York, Marjorie was a US citizen. Her father, David Weaver, was a decorated US military chaplain, having served with valor in the Korean War. Marjorie's mother, Clara, was the church organist during the times when David pastored churches. Clara also loved to read. It was said that she often took home from the library a suitcase full of books at a time – reading all of them before the due date.

David and Clara Weaver.

Chaplain David Weaver.

The second of the four Weaver children, gentle and pretty Marjorie graduated as a registered nurse from Kentucky Baptist Hospital School of Nursing in Louisville, before completing her college education at Georgetown College, Kentucky. She admitted that she was known among her classmates as a practical joker, but because of her sweet disposition none of the faculty ever suspected her.

Hendon Jr. and Marjorie married on her birthday in fall 1940. She was 23 and he was 24. Within days of their marriage the US government started ordering American missionary families, including Hendon Jr.'s mother and sister, Florence and Cita Harris, out of China. Therefore, the young Harris couple's door to service in China at that time slammed shut. However, they presumed that it would open again.

Several years later when I was born, I was named after the elderly lady, Charlotte Reid, whom my father had asked to pray that he would find an appropriate wife. My father was handsome and charming so had no problem finding girlfriends, but since he wanted to go to China, he needed a wife who also had that call. Not every young lady of his era was eager to move to Japanese occupied China.

Hendon Jr. and Marjorie three days before their wedding.

Later my parents found out that Charlotte Reid did not just pray for a wife for Hendon Jr.; she prayed specifically that he would meet and marry my mother who had already declared that she wanted to go to China as a missionary. Unknown to Charlotte Reid was the fact that at the time she started praying Marjorie was still engaged to someone else.

Marjorie grew up in a very strict household. After she graduated from nursing school she took a job while still living in her parents' home. At that time her father required her to turn her entire paycheck over to him. Out of her paycheck she received an allowance. Her father was a good, God fearing man but quite controlling.

Rufus Gray, my father's seminary classmate who was accepted as a missionary by the Southern Baptists in 1940 when my father was passed over, was in language study in the Philippines at the time the Japanese bombed Pearl Harbor. A hobby of Rufus was photography, but the Japanese presumed that he was using his camera to spy. Rufus was tortured to death by the Japanese in the prison camp which continued holding his wife and infant son. It was providential that Hendon Jr. did not get to go to China along with Rufus.

After marriage Hendon Jr. initially pastored a church not far from Louisville in southern Indiana. Marjorie Florence, their first child, was born while they lived there. On the way to the hospital to deliver the baby both (mother) Marjorie's father and husband were in the car and got into an argument about how fast to drive. With Marjorie in advanced labor, Hendon Jr. stopped the car and relinquished the steering wheel to his father-in-law.[407]

Next the family moved to the Chicago area so that Hendon Jr. could earn his doctorate in theology from Northern Baptist Seminary – which he completed in 1945. His over 450 page doctoral dissertation titled "A Short History of Religious Liberty" is still timely today. Aunt Cita recalls visiting her brother one time while he was finishing his thesis. Cita stated that as soon as Hendon wrote a page Marjorie would type it –sometimes staying up all night.[408]

While in the Chicago area Hendon Jr. and Marjorie started four churches – one at a time. He was the preacher, but Marjorie was also an excellent Bible teacher and demonstrated love for all types of people. She never met a stranger. It seemed that all her life everyone she met was an instant friend.

One unfamiliar with evangelists might wonder why they are so dedicated in trying to win others to Christ. The reason is that the Bible teaches that unless any person accepts Christ as savior, he or she is doomed to eternal punishment in hell. Just as a fireman is adamant about wanting to rescue anyone in a fire, so an evangelist, out of compassion and based on strongly held convictions, is also trying to rescue people from eternal flames.

Perhaps one could also compare the evangelists' passion to someone who found a cure for cancer. Would not that person with the cure for cancer want to tell everyone – especially if he could offer it as a free gift? However, to accept that "remedy" one would have to forsake the mud packs or whatever other cure he had been trying. The missionaries believed what Jesus said: "I am the way, the truth, and the life. No one comes to the Father except through me."[409]

The paradox was that those same missionaries believed in freedom of religion, that each person had to make his own choice – good or bad. However, how could one make an informed choice about Christianity if he did not know that option was available?

My parents supported themselves and their church planting ministry in the Chicago area through both Marjorie's wages as a registered nurse and by buying houses, first fixing them up while living in them, and then re-selling them for a profit – which is known today as "flipping a house."

During that period of time in the Chicago area Lillian, then I, and then Hendon III, were born. The year I (their third child) entered the world, I almost died from pulmonary problems. During that cold Chicago winter our family was living in a house without interior walls and without adequate heat. Fortunately, that house was soon fixed up and sold. During that time Hendon Jr. and Marjorie learned the lesson that William Powell, his grandfather, espoused many years earlier: "God will provide."

My parents both still yearned to go to China as missionaries. However, by then the political situation in China was still volatile with the continuing civil war between the Nationalists and Communists. About then (the late 1940's), my Harris grandparents left China for the final time – just before the Communists took over their area. The foreign missionaries who stayed in China after Communist takeover were expelled or killed.

Four Harris children shortly before going to Taiwan – Lillian, Hendon III, Charlotte (in front), Marjorie Florence.

Among the four churches that Hendon Jr. and Marjorie started in the Chicago area at that time were Donald Smith Memorial Baptist Church in Oak Lawn, Illinois and Airport Baptist in Chicago (now Emmanuel Baptist in Hinsdale).[410].

Donald Smith Memorial Baptist was named after a young man who drowned during the early years of that church. On hearing of the tragedy Hendon Jr. immediately went to the pond where police officers standing on the bank were thrusting poles into the water trying to locate the boy. With all his clothes on the young preacher dove right in and brought the young man to the surface on his first try, but it was too late. The boy was already dead. No one was able to resuscitate him.

Airport Baptist was named because of its location near Chicago's Midway Airport. Years later when my parents were overseas I spent one summer break from college working in Chicago while I stayed in the home of Howard and Florence Voegtlin, members of Airport Baptist.

Earlier my older sisters, missionary kids with no Stateside home, also received the Voegtlin's hospitality. Later widow Ruth Price of Beverly Emmanuel Baptist Church housed us.

Florence Voegtlin reconfirmed a story that my parents had told me: As Airport Baptist was being founded one icy day my parents were canvasing house to house in their neighborhood. My mother slipped on the ice and sat down hard, injuring her coccyx (tailbone). Florence invited my parents into her home.

Once in the house my parents told them of Jesus' love for them. The Voegtlins became believers in Christ. The Voegtlins and their three sons Donald, Roger, and Ken joined the church.

Donald and Roger both later went into full time ministry. Roger married Sharon Binkley, also from Airport Baptist. Their ministry has been extensive, positively affecting the lives of thousands of individuals. Roger and Sharon's daughter and her husband are now foreign missionaries and their son is a pastor. Unfortunately my mother suffered with her injured coccyx for the rest of her life. However, she would have said that her injury was worth the saving of even one soul.

For years the US Government and many American individuals had supported the Nationalists in China with enormous amounts of aid. However, when the Nationalists fled to Taiwan it was presumed to be a losing battle. People believed that with Taiwan only 112 miles (180 kilometers) off the coast of China surely the Communists would soon over run them.

However, Hendon Jr., whose heart "beat in Chinese" was not ready to forsake his Chinese brothers in Taiwan. When the United States seemingly turned its back on the Nationalist Chinese who had fled to Taiwan, Hendon Jr. personally visited 96 American senators in Washington, DC pleading for support of the Nationalists.[411]

Hendon Jr. at that time also corresponded with U.S. Lieutenant Generals A. C. Wedemeyer and Claire Chennault proposing that the Americans continue to support the Nationalists and offering his services to preach to the troops in Taiwan.

My father received two lengthy letters – one from Claire Chennault and another dated 14 November 1949 from General A. C. Wedemeyer on letterhead of the Headquarters of the Sixth Army, Office of the Commanding General. Wedemeyer offered encouragement to the plan that my father had proposed and stated:

> The motivation of your ideas is a commendable one, and one which I realize emanates from a sincere desire to alleviate the present deplorable situation in China before it is irrevocably lost. It is my considered opinion that the initiative in the struggle of democratic versus communistic elements in China should be maintained by the United States....
>
> Further, your plan would carry inestimable value in siding the general international picture by the assertion that the American people ideologically now (and perhaps later, materially) intend to oppose the Communist powers in China and in fact everywhere and to do all possible to aid in the resurgence of democratic forces.

HEADQUARTERS SIXTH ARMY
OFFICE OF THE COMMANDING GENERAL
PRESIDIO OF SAN FRANCISCO, CALIFORNIA

14 November 1949.

Dear Reverend Harris:

Please accept my apologies for delay in answering your letter, but my time has not been my own since assuming command of the Sixth Army recently. Your letter was indeed a provocative one. I am in full agreement with the plans which you and Captain Bundy have outlined for an American organization dedicated to the best interests of the Chinese people. However, I am sure you will understand my position in not being able to provide unequivocal support to the program. As you know, I am a soldier and as such it is incumbent upon me to implement policies, not to formulate them. For that reason, I regretfully must decline to take an active part in the organization you visualize.

I would like, however, to offer my encouragement and the hope that you will continue to sponsor the lofty ideas expressed in your proposed organization. The motivation of your ideas is a commendable one, and one which I realize emanates from a sincere desire to alleviate the present deplorable situation in China before it is irrevocably lost. It is my considered opinion that the initiative in the struggle of democratic versus communistic elements in China should be maintained by the United States, just as it should be in the intricate international problems in other areas of the world. Also other nations with parallel or similar aims to ours should integrate and coordinate their efforts with ours in order to further their realization.

An organization such as yours would, I feel, accomplish a great deal in stimulating awareness in the minds of Americans not only of the tragic eclipse of democratic China, but the complete usurpation of individual rights and freedoms by the rising Communist power. In turn, any activity of the citizens of our country aimed, whether nationally or morally, against the Communist aggression, would be a decided factor in raising the morale of the oppressed Chinese.

Further, your plan would carry inestimable value in aiding the general international picture by the assertion that the American people ideologically now (and perhaps later, materially) intend to oppose the Communist powers in China and in fact everywhere and to do all possible to aid in the resurgence of democratic forces.

I hope the above information will be of some assistance to you.

Faithfully yours,

A. C. WEDEMEYER,
Lieutenant General, USA
Commanding.

Reverend Herdon Harris,
The New Testament Evangelization Society,
312 South LaGrange Road,
LaGrange, Illinois.

文通部民用航空局直轄空運隊

CIVIL AIR TRANSPORT
KAI TAK AIRFIELD
HONG KONG

IN REPLY PLEASE QUOTE
OUR REF. NO. ....

CABLE ADDRESS
CLAULT HONGKONG

30 December 1949

Dr. Hendon M. Harris, Th.D.
The New Testament Evangelization Society
312 South La Grange Road
La Grange, Illinois

Dear Dr. Harris:

Your letter dated October 28th just arrived, and I appreciate your sentiments very much. I feel as you do that an American force of officers and technical specialists should be organized to train and supervise the Chinese in their war of resistance against Communism. For more than two years I have advocated the organization of a small but very effective international air unit to support the Chinese armies in the field. I spent more than five months in Washington at my own expense during the past summer, and I discussed this project with officials of the Department of National Defense, the State Department and members of Congress. Nearly all except the State Department approved the idea. It would be impossible to accomplish anything of decisive value without the approval of the State Department because of the restrictions which could be placed upon American citizens and upon the export of war materials by that Department. You probably know that our late President, Franklin D. Roosevelt, unofficially approved the A.V.G. and issued confidential instructions to War, Navy and State to permit personnel to volunteer for service in that unit and to permit the export of airplanes, engines, guns, bombs and other equipment needed by the unit in battle. Had we not received this sort of support, the A.V.G. could not have flown in combat.

There are a number of societies in the United States which support active American resistance to Communism in China and the Far East, and I do not believe that the organization of an additional society would accomplish anything unless it could be followed up by the organization of supervisory army units and a small air task force. The Chinese Reds are actively supported in this same manner by Russians. Many Red units have been trained by Japanese in the use of Japanese equipment and Russian engineers supervise the movement of their supplies and replacements. So far as I know, no Russian actually fires a gun but the planning and logistical organization is all Russian-managed.

We should not hesitate to employ the same devices in support of our friends, the non-Communist Chinese and other Asiatic peoples. It seems that little can be done in a material way unless you good people of the United States convince the President and the State Department that it is to the interest of our country to oppose the Russians in their attempt to communize Asia. Since we have a democratic form of government, for which we are all duly grateful, we must follow the methods of a democracy. In this case, the majority of our people must convince our top-level leaders that we desire to fight Communism in Asia as well as in Europe.

Again thanking you for the sentiments expressed in your letter and wishing you every success in your effort to aid non-Communist China, I am

Most sincerely yours,

C. L. Chennault

C. L. Chennault

Invitation to Go to Taiwan

Chenault's letter dated December 30, 1949 was followed by a Western Union telegram from him to my father.

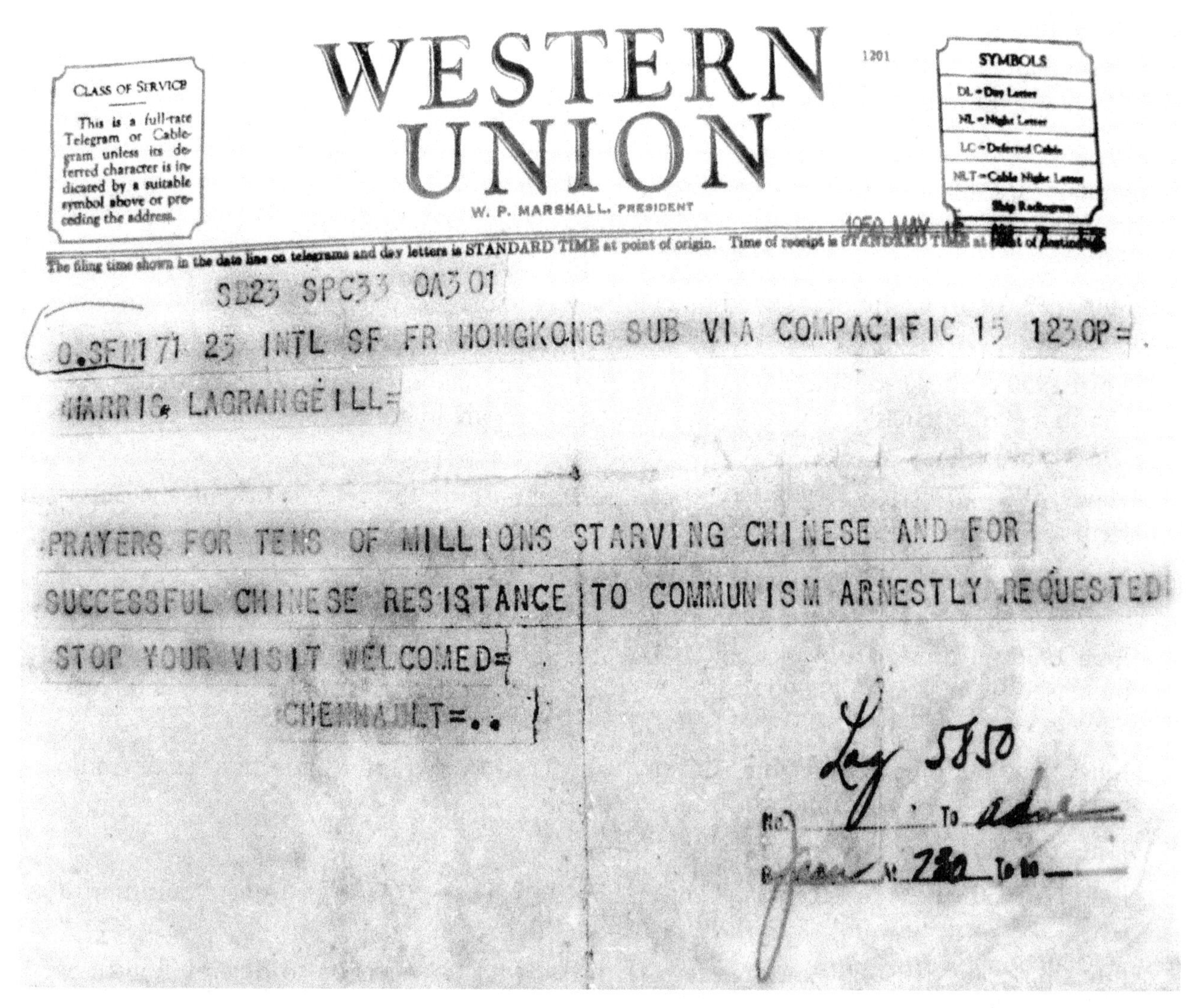

CLASS OF SERVICE

This is a full-rate Telegram or Cablegram unless its deferred character is indicated by a suitable symbol above or preceding the address.

WESTERN UNION

W. P. MARSHALL, PRESIDENT

1201

SYMBOLS

DL = Day Letter

NL = Night Letter

LC = Deferred Cable

NLT = Cable Night Letter

Ship Radiogram

1950 MAY 15

The filing time shown in the date line on telegrams and day letters is STANDARD TIME at point of origin. Time of receipt is STANDARD TIME at point of destination

SB23 SPC33 OA301

O.SFN171 25 INTL SF FR HONGKONG SUB VIA COMPACIFIC 15 1230P=

HARRIS LAGRANGE ILL=

PRAYERS FOR TENS OF MILLIONS STARVING CHINESE AND FOR SUCCESSFUL CHINESE RESISTANCE TO COMMUNISM ARNESTLY REQUESTED STOP YOUR VISIT WELCOMED=

CHENNAULT=..

Log 5850

This telegram reads: PRAYERS FOR TENS OF MILLIONS STARVING CHINESE AND FOR SUCCESSFUL CHINESE RESISTANCE TO COMMUNISM ARNESTLY [SIC] REQUESTED STOP YOUR VISIT WELCOMED. CHENNAULT.

Chennault, as spokesperson for Chiang Kai Shek, had invited Hendon Jr. to go to Taiwan. When my father left for that country he took with him 20 thousand Bibles and 70,000 Gospel booklets to present to Madame Chiang Kai Shek (wife of the leader of the Nationalists) for distribution to the soldiers and officers of the three military divisions in Taiwan.[412]

Timeline

1949 - Nationalist Chinese troops fled from China's mainland to island of Taiwan.

Fall 1949 - Concerns arose among a few US Congressmen[413] and also in Hendon Harris Jr. regarding whether the US would protect the Chinese in Taiwan. Harris wrote letters.

November 14, 1949 - General Wedemeyer replied to Hendon Harris Jr.'s letter regarding defense of Taiwan. He was sympathetic and encouraging but was not able to help.

December 30, 1949 - Claire Chennault sent letter in response to Hendon Harris Jr.

January 5, 1950 – Truman stated that the US would not interfere in whatever happened in Taiwan.[414] Knowland and a few others in Congress argued for months for defense of Taiwan.

March 22, 1950 - CIA report stated that China would probably seize Taiwan between June and December 1950.[415]

April 1950 - After receiving telegram invitation from Claire Chennault to go to Taiwan, Hendon Harris Jr. left his pastorate in Chicago to prepare to go. He took 20,000 Bibles and 70,000 tracts.

May 9, 1950 – Harris wrote letters to congressmen asking for defense of Taiwan. Harris went to Washington and met with over 90 congressmen, urging defense of Taiwan.

May 1950 - Harris arrived in Taiwan and preached until that fall. 38,000 prayed to accept Christ.

June 1950 - China's army positioned directly across the 112 mile wide Taiwan Strait increased from around 40,000 to 156,000 backed by 300,000 additional troops and some Soviets.[416]

June 21, 1950 - per letter from Taiwan Broadcasting Station Harris had made radio broadcasts to US and mainland China.

June 25, 1950 - Korean War broke out.

June 27, 1950 - President Truman sent 7th Fleet to Taiwan Strait to prevent a Communist advance toward Taiwan.[417]

June 28, 1950 - Senator H. Alexander Smith declared US decision to defend Taiwan as "God guided."[418]

July 20, 1950 - Eugene, younger brother of Hendon Harris Jr., killed in Korean Conflict. Family was notified that fall.

1951- Hendon Harris Jr. family moved to Taiwan.

Summer 1953 – After Korean War truce Hendon Harris Jr. preached to 13,500 Chinese POWs in Pan Mun Jom, Korea. He also encouraged POWs to go to Taiwan – not to return to China.

January 23, 1954 – 14,000 Chinese Prisoners of War from Korea were welcomed to Taiwan.

In his book *The Asiatic Fathers of America* Hendon Jr. wrote in the third person:

> In 1950 when he [Hendon Jr.] lived at Chicago he used to go daily to the forest preserves to pray for Taiwan, which was unprotected and expected to be invaded by the Communists in six weeks. Feeling that the Chinese needed his help, he went to the Congressmen in Washington and pleaded with them not to abandon the Free Chinese.

Then, with only enough money for a one-way ticket (and an invitation from General Claire Chennault):[419]

> ...he left his wife and four children to go on a seemingly ridiculous one-man expedition to help the Chinese in Formosa [Taiwan].
>
> ...Madame Chiang [Kai-Shek] was touched by his arrival... [Harris] was invited to speak to the Chinese Armies. In his meetings... more than 38,000 Chinese became Christians during that... spring and summer. (Shortly afterwards) the Communists invaded Korea and Truman sent the Fleet to protect Taiwan.[420]

Hendon Jr. meeting with Madame Chiang Kai Shek in Taiwan.

The news article below was written by Spencer Moosa for the Associated Press shortly after Hendon Jr.'s arrival. Reportedly Moosa, as the last foreign journalist to leave the Chinese mainland, was on one of the final four planes reserved for Nationalist leaders headed to Taiwan.[421] Therefore, he was in Taiwan when my father was there so definitely knew what was happening.

Morale High in Formosa

U.S. Minister Wins Converts, Asserts Island Will Not Fall

By SPENCER MOOSA

TAIPEI, Formosa (AP)—Late last April the Rev. Hendon M. Harris was driving in the Hidden Pond Woods forest preserve near Chicago. He felt sleepy and halted his car to take a nap.

He was awakened, he recalls by a strong wind and looking at the sky saw a bear and dragon fleeing from pursuit by a wounded lamb.

His vision, he decided, was God's way of vouchsafing that Communism would go down in defeat —and God's way of telling him to join in the fight.

Harris is pastor of the Airport Baptist Church in Chicago. His congregation was raising funds to build a new church. It decided to use the money to send him to Nationalist China.

BROUGHT MESSAGE

Harris flew to Formosa. He arrived in May, when Nationalist fortunes were at very low ebb and their cause was lost in the eyes of the world.

Harris brought a message of hope. The story of his vision spread from one end of the island to the other.

Since his arrival he has spoken before Nationalist troops everywhere. His evangelism has won 24,000 converts to Christianity.

Included among the converts was an entire group of 7,000 Nationalist combat troops who heard him speak.

Harris describes the morale of Nationalist troops as excellent and says they are anxious to prove themselves in battle.

WON'T FALL

"They represent 600,000 reasons why this island will never fall," asserts.

To the pastor's way of thinking, the events on the mainland which led to the Communist conquest were all part of a divine plan.

As part of this—in his words—"the Supreme Being provided the Chinese leaders with a refuge here in Formosa to cleanse themselves of corruption."

He predicts that the Nationalists will return to the mainland, and that the people, as result of suffering what he calls the Red scourge, will turn in large numbers to Christianity.

NO FANATIC

Harris in 34 years old. There is nothing about him to suggest fanaticism. He was born in China, the son of missionaries.

He is director of the New Testament Evangelization Society of LaGrange, Ill., where his wife and four children live.

Harris says he is in Formosa not only as a Christian but as a patriotic American, adding:

"Communism is just as much a threat to the people of America as it is to people of Nationalist China and other free countries."

Faith in Formosa

Last May the Rev. Hendon M. Harris, above, pastor of the Airport Baptist Church in Chicago, flew to Formosa. Since his arrival he has spoken before Chinese Nationalist troops everywhere. His evangelism has won 24,000 converts to Christianity. Harris describes the morale of Nationalist troops as excellent and says "they represent 600,000 reasons why this island will never fall." (Wide World Photo)

As Hendon Jr. continued preaching the number of converts continued rising. Another 1950 news article written later that same year by Rev. John Evans (who wrote for *Chicago Daily Tribune*) is titled "Chicago Pastor Carries Cross to Formosans: Sees Rebirth of China in His Work." Evans stated by that time the number of converts in Taiwan had risen to 34,000. Eventually it reportedly reached 38,000.[422]

CHICAGO PASTOR CARRIES CROSS TO FORMOSANS

Sees Rebirth of China in His Work

BY THE REV. JOHN EVANS

Until last April the Rev. Hendon M. Harris, 34, was the pastor of the Airport Baptist church, a new congregation which meets in a portable school house at 64th st., and Tripp av. All was going well and still is, but the Rev. Mr. Harris has been in Formosa since last May and meanwhile has been leading his flock from a distance of 8,000 miles.

Many things happened in a few days last spring to send him back to China where he was born in a missionary family. One was a vision, he told friends, which came to him while he was resting in the forest preserve. Nearly asleep, a strong wind sprang up which awakened him. In the sky he saw images of a bear and a dragon fleeing from a wounded lamb.

Feels Call to Formosa

A wounded lamb is a symbol of Christ, the bear a symbol of Russia and the dragon, China. He felt that this was God's way of telling him how communism would be defeated and that he was needed in Formosa to convert Chinese nationalist soldiers to Christianity. He and his parents have been associated with Gen. Claire Chennault. The general told the Rev. Mr. Harris that an evangelistic campaign over a few months among the nationalists on Formosa would be desirable.

The Rev. Dr. Robert T. Ketcham, national representative of the General Association of Regular Baptists, 155 N. Clark st., with which the Airport church is affiliated, related the Chennault episode, adding that with Gen. Chennault's backing in Washington, permission was granted by the state department for Harris to visit Formosa.

The Rev. Mr. Harris asked his congregation how money might be raised. Arthur B. Leslie, 6142 S. Tripp av., clerk of the Airport congregation, said the church had been saving for a new building and had collected $1,700. The congregation voted to give him this money.

He Preaches Hope

Harris took off by air for Formosa. He arrived in May when the nationalist cause seemed lost. He preached hope. The story of his vision spread thru the island. More than 7,000 nationalist combat troops were converted to Christianity. The movement spread with lightning rapidity. Leslie said the number now is 24,000 converts. Dr. Ketcham, in Ithaca, N. Y., said 34,000.

Messages from the Rev. Mr. Harris to his congregation, transmitted thru Mrs. Harris, 312 S. LaGrange rd., La Grange, state that the morale of nationalist troops is excellent, and that they are anxious to try their mettle in battle.

"They represent 600,000 reasons why this island will not fall," Mr. Harris stated. "God provided Chinese leaders with a refuge here in Formosa to cleanse themselves of corruption."

The Rev. Dr. Eugene M. Harrison, of Wheaton college, is supplying the pulpit of Airport church during the Rev. Mr. Harris' ab-

A Chinese newspaper reported at that time Father "met with 60 military leaders of the Nationalist Party's Army, and he gave them much encouragement."

Below are photos from my parents' album of my father preaching to large groups of Chinese troops. Father stated that after he preached to troops, repeatedly many would rush forward to pray to accept Christ while others stood back and laughed at them.

For years in China the Chinese Nationalists had been supportive of Christian missionaries, while the Communists opposed Christianity, often killing the Christians. At that time the Communists opposed religious freedom even for Buddhists. Hendon Jr.'s doctoral thesis had been about the history of religious freedom, an ideal of which he was a staunch supporter, even if the other person's choice of religion was not Christianity.

Soldiers in Taiwan raising their hands signifying that they wanted to accept Christ. Hendon Jr. is second from left in foreground with back to camera.

More soldiers praying with Harris to accept Christ.

Some of the Nationalist troops and their families who moved to Taiwan in 1949 already had become Christians. Some had served in China as Christian leaders.

The recent biography of the retired Chief of the Asian Division Library of Congress, Dr. Hwa-Wei Lee, reports that his father was among the Nationalists who, with his family, moved to Taiwan at that time. Before joining the military Lee's father had been a professor then the principal of a Christian school in Fujian, China. In Taiwan his parents were active in their faith.[423]

Those Chinese Christians, along with the thousands of new converts, had an effect on the general populace of that small island. Among other things, people in Taiwan began commonly using "*li bai tian*" (literally "worship day") for "Sunday." Every day of the week that followed was a count down from worship day. Monday was *li bai yi.* (worship day plus 1), Tuesday was *li bai er* (worship day plus 2), etc. (Just a few years ago while visiting China I used "*li bai tian*" to refer to Sunday. The people on whom I was testing my rusty Mandarin smiled – as one would do if I used the outdated word "groovy" in English. They said that they immediately knew that I had learned Mandarin in Taiwan.)

It had to have been an act of God that the Communists never invaded Taiwan to take the Nationalists' last stronghold – especially since the Nationalists, believing that they were the true China, had taken China's national treasures. Those treasures are to this day in museums in Taiwan.

After the Nationalist Army and many others fled from mainland China to Taiwan in 1949, a few female Southern Baptist Missionaries who had been serving in China followed them there. One, Ola Lea, had been an instructor of women in Hendon Sr.'s, seminary in Kaifeng.

Favorable mention about Hendon Jr. and his initial 1950 trip to Taiwan is found in the Southern Baptist 1951 *Annual* (reporting on 1950) in an article by missionary Bertha Smith:

> The Lord knowing our need for a missionary pastor, sent Dr. Hendon Harris Jr., son of China missionaries, out on a summer preaching mission. Although he came especially to preach to China's armed forces, he gave our church a few weeks, and from his first week's meeting the church was filled and has continued so.[424]

That Southern Baptist report mentioned individual conversations that Hendon Jr. had in leading people to Christ. It also stated that "nearly four million of the nine million people of Formosa are from the various mainland provinces – refugees from communism."[425]

While in Taiwan Hendon Jr. also spoke on Taiwanese television in broadcasts to the United States asking American citizens to assist in Free China's opposition to the Communists. His goal was to maintain religious freedom for Chinese.

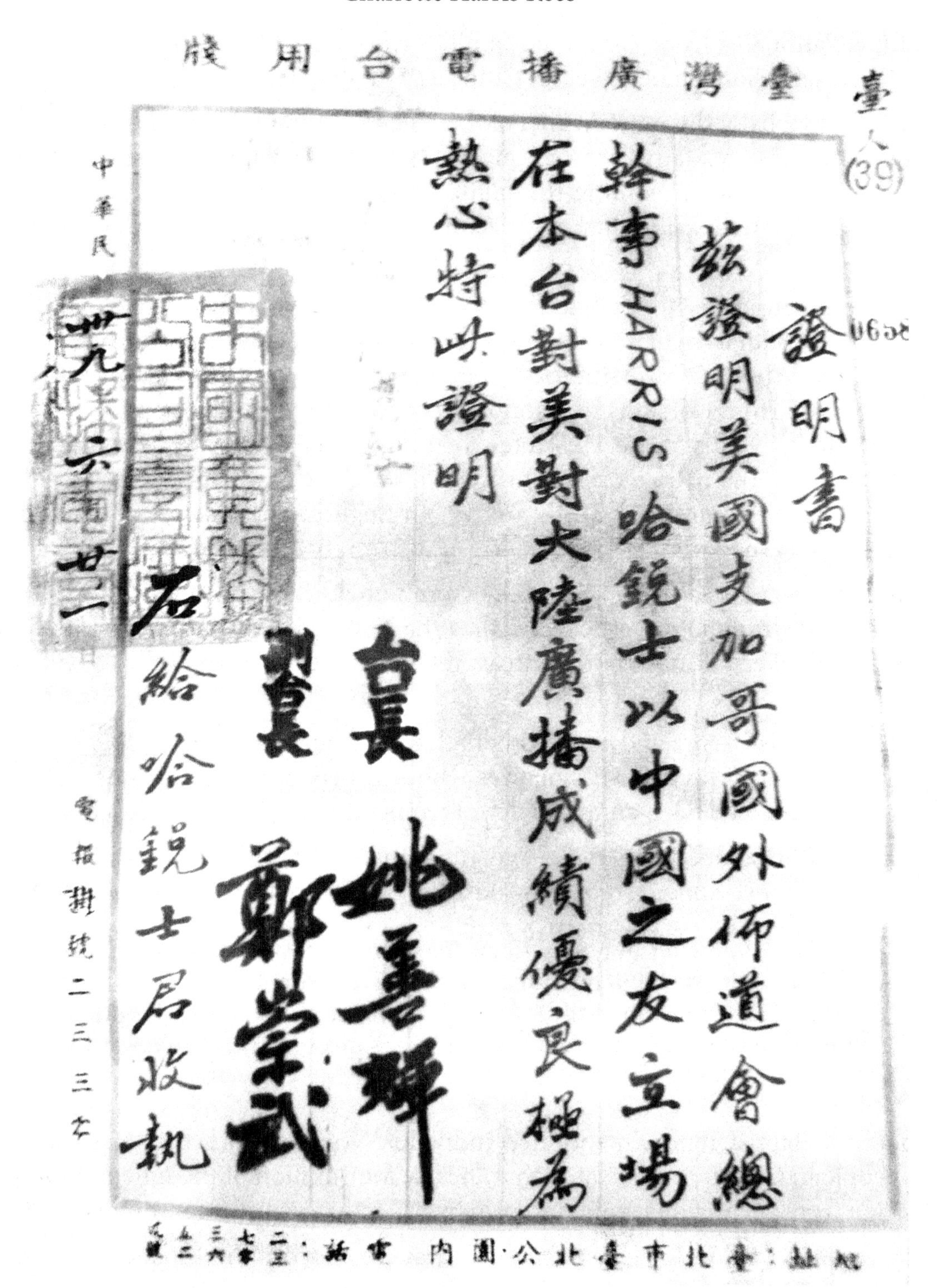

臺灣廣播電台用牋

臺人(39)

0658

證明書

茲證明美國支加哥國外佈道會總幹事HARRIS哈銳士以中國之友立場在本台對美對大陸廣播成績優良極為熱心特此證明

台長 姚善輝

副台長 鄭崇武

石 給哈銳士君收執

中華民國卅九年六月廿一

電報掛號二三三[illegible]

地址：臺北市臺北公園內 電話：[illegible]

Taiwan Broadcasting station statement dated June 21, 1950 indicating that Harris spoke via their station to America and mainland China.

The Harris family album has several letters from various Chinese leaders praising my father's early efforts in Taiwan.

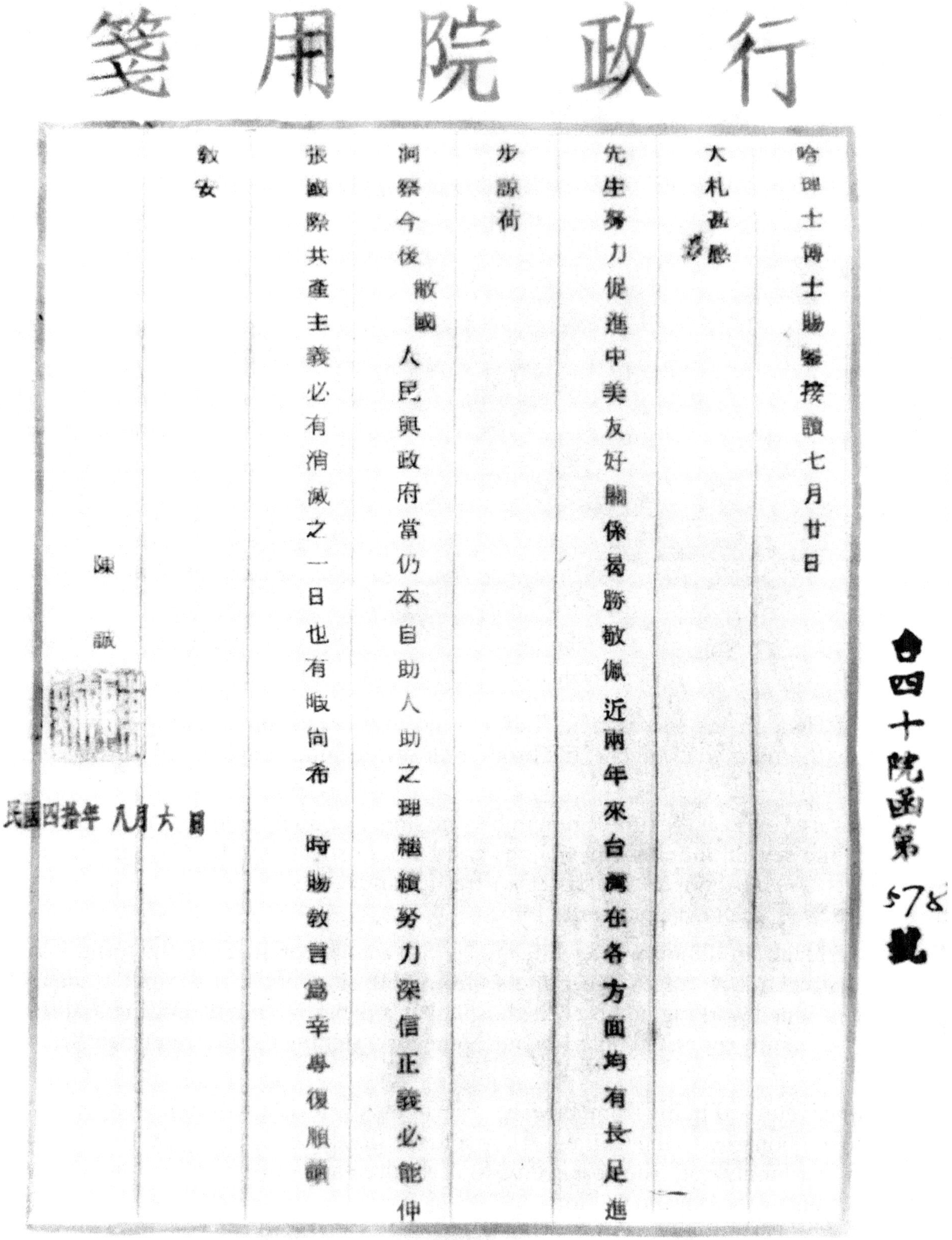

行政院用箋

哈理士博士勛鑒接讀七月廿日

大札敬悉

先生努力促進中美友好關係曷勝敬佩近兩年來台灣在各方面均有長足進

步諒荷

洞察今後 敝國人民與政府當仍本自助人助之理繼續努力深信正義必能伸

張國際共產主義必有消滅之一日也有暇尚希 時賜教言爲幸專復順頌

教安

陳誠

民國四拾年八月六日

台四十院函第578號

Thanks from the Chinese in 1951.

Words of thanks to Dr. Hendon M. Harris Jr.

The joy of Hendon Jr.'s successes in Taiwan that year was soon co-mingled with sorrow brought by the news that Eugene, his brother closest to him in age, had been killed in the Korean conflict.

Although very young at the time, I remember missing my father terribly as his trip to Taiwan stretched into several months. My siblings and I would sneak into his dresser and take out undershirts. We all thought that his clean fresh laundry smelled like him. Snuggling into those shirts we somehow felt closer to him.

Our last residence in Illinois was a white house with a large porch across the front. It sat on a tree-lined street in LaGrange. Tree branches from both sides of the street met in the middle. Mother continued working nights at the hospital as a registered nurse. During that time she gave another family room and board in our home for watching us children when she worked or slept.

When father returned home from Taiwan, my parents decided it was time to carry out their plan to be missionaries to the Chinese. They sold the house and most of our belongings then gave away our family dog. They packed us four children and our few remaining possessions into the car and drove west to California.

Father never applied again to the Southern Baptist mission board, which would have provided a salary with benefits. Hendon Jr. and Marjorie formed their own organization, the Baptist

Evangelization Society (later known as the Hundred Nations Crusade). On the trip west across the United States they raised pledges for missionary support. On reaching California, Father went ahead to prepare a place for us in Taiwan.

At that point my parents did not have enough money for passage to the Orient for us all. Consequently Mom worked as a nurse for a few weeks in San Francisco until she earned the rest of the fare. She then booked passage for us on a freighter since that was less expensive than a regular passenger ship.

After the ship's arrival in Japan, we flew to Taiwan. My mother cabled ahead to tell Father our arrival time, but for some reason he was late to the airport. Mom was in a country where she did not speak the language; she had no money left; and she had four young children at her side. For a few minutes there was consternation until Father arrived.

At Home in Taiwan (Also Known as Formosa)

On moving to Taiwan Father left behind the meek and mild life, and so did the rest of us. But what would one expect of a man whose grandfather and father had lived such unusual lives? What would one expect of a man who in his childhood had lived among warlords and traveled the world? Below is a poem that he wrote.

"Rebellion"

Laughter and Tears[426]
By Hendon Harris Jr.

Do you expect to make of me
A person well content,
When yesterday my eyes did see
A distant battlement,
And, far across the summer lea
Brave knights to battle bent?

What is the world that you should hide me,
Like a suckling child?
I, too, 'mid heroes fair would ride
With heart as proud and wild,
Leaving behind, securely tied
A Life that's meek and mild.

Our arrival in a small village in Taiwan could not have been more electrifying to the villagers than if we had green skin, blue hair, and arrived in a spaceship. At that time no one in the village had a television, computer, nor had they ever before seen Caucasian, blue-eyed, blond (some of us) children. We had non-stop attention. As a typical five year old, I remember jumping around or making faces just to hear the sighs and giggles.

A tall bamboo fence surrounded our yard. One at a time three different bamboo fences were knocked over by curious villagers of all ages peeking through the cracks in that enclosure. Finally a tall concrete wall with broken glass embedded in the top had to be built around our yard to give us privacy. The final walls were similar to those of the compound in China within which Father lived as a child.

This building housed both our residence and the orphanage. Note bamboo fence in background through which numerous people gawked at our family. Front L to R Charlotte, Lillian, Back Marjorie Florence, Hendon III, Marjorie Harris.

Playing on their curiosity, whenever Father wanted to gather a crowd he would give his offspring drums, tambourines, and cymbals. We would march down the street. When enough people had gathered, we children would sing Chinese songs about God that Father had taught us. Then he would preach in their native tongue. From the ensuing nucleus of believers he soon started a church.

Our parents taught us to love God and others. As a young child, I myself prayed to accept Christ and was baptized with other recent converts in the river. I was so small that I had to be carried out into the swift current.

Monkeys and a flying squirrel were brought home as pets for us and we took rides on the backs of water buffalo. Father insisted that we learn to eat strange and exotic foods and respect Chinese culture.

Father told us wonderful bedtime stories, wrote songs, a novel, books of poetry, and a cantata in addition to his missionary work. Later he even produced a movie. The creativity and wonderful sense of humor of both of our parents enriched our lives. Two more children were born to my parents while we were in Taiwan—bringing the running total to six.

Just as people in China had racial slurs for Caucasians, some in Taiwan, whose population largely came from China, used those same slurs on us as well.

Our town was not far from the city of Taipei. One day my father decided to play a trick on our tormentors, whose racial slurs for Caucasians included *da bizi* (big nose).

Father had a huge fake nose attached to glasses and wore it as he drove through Taipei. Having never seen such a device before, people stared and pointed. Further down the street he turned the jeep around. Putting the fake nose on my sister, Marjorie, he drove back so the same people could see it on her.

At another location he wore the fake nose and pretended to have car trouble. When he looked under the hood people gathered around. Not knowing that my father spoke Mandarin and could understand what they were saying, a big debate followed among the spectators about whether the nose was real or fake. After a while my father said in Mandarin, "Can I help it if I was born this way?" Then when he saw that the people were embarrassed, he took off the fake nose and everyone laughed.

Mom also had a great sense of humor, which one needs when learning a foreign language. We all laughed when one day she excitedly told us in Chinese that a train (derived from the Chinese words "fire" and "vehicle") was in our yard when she was trying to say that a neighbor's escaped turkey was in the yard. (Turkey is formed from Chinese words for "fire" and "bird.")

The first time Mom tried to pray in Chinese in public she used the word for "America" in place of the word for "devil" as she prayed: "Lord protect us from the wiles of America."

"**What** were you trying to say?" Father asked.

Fortunately she was later able to laugh at herself. Who knows what the Chinese thought!

All of us children were young enough that learning to speak Mandarin came easily. One day when an important Chinese man came to see my dad only my mother and three-year-old Hendon III were home. The man spoke so rapidly that my mother turned to Hendon III for help.

Hendon III was very proud that he had just started Chinese nursery school. After he had acted as the interpreter he turned to the man and said in words which my mother also understood, "Please forgive my mother. She never went to school." (Meaning that she never attended a Chinese school like Hendon III had.)

Father was literate in several languages. From my earliest memories, he always had a study in our home and spent hours among his numerous books and stacks of paper for his research and sermon preparation.

One of father's songs shown below was inspired by Psalm 139 and was written in those early years when a missionary friend was leaving Taiwan. Note Father's Chinese perspective – that one might be more lonely to leave Taiwan than to stay.

"Wings of the Morning"

Words and music by Hendon Harris Jr.

Though I take the wings of the morning
And fly across the sea
Even there the Lord will guide me
And keep me company

Farewell till some bright tomorrow
Farewell you children of men
The sheep of God's flock
Will be scattered apart
But in Heaven we'll all meet again

Our houseguests ranged from Chinese generals to mountain aborigines, who were just a few years removed from being headhunters. One of our succession of amahs (nannies) over the years had tiny bound feet. Occasionally our parents took us on trips up into the mountains of Taiwan.

The sights, sounds, tastes, smells, and memories of our childhood are burned in our minds and hearts. Taiwan in that era was very different from America, but I have never regretted my years in Asia. I feel that my life is richer and fuller for having lived there. It seems that once immersed in Chinese culture, one never goes away without leaving a piece of his heart there.

All those years Father was an independent missionary so had no regular paycheck. The ministry was supported by contributions from churches and individuals and by Father's ingenuity. To supplement the contributions, he frequently bought antiques in the Orient and sold them when he went to the United States.

"Let Me Play God's Game"

By Hendon Harris Jr.

Let me play god's game along the Road
That stretches around the bend
From Hokkaido [Japan] to the Taj Mahal
Until my journey's end.

Courage, Endurance, Sacrifice

Let me play God's game along the fence
When the Bear [Communism] grins in disdain
And do my share for my Eastern friends
In the sun or in the rain.

Let me play God's game with the golden men
Who trust in our mutual Lord
As long as I live, let me go and give
The message that's in the Word!

Let me play God' game with all I have
And turn my pockets out
To lay the last dime on the line
Without a sigh or doubt

Let me play God's game in joy or grief
While hopes and planes still fly.
And when I'm through, its in the East
That I at last, would lie.

The poem below was written by Dad just months after the family arrived in Taiwan and after he had just left on another trip. It was a strange new world that we were making home. No one else in our village aside from us spoke English and we (except for Dad) were just learning Chinese. The thought that he was leaving us there must have been very unsettling to me.

"Charlotte"

Laughter and Tears[427]
By Hendon Harris Jr.
London, Feb. 1952

Charlotte, my little girl
Only five brief years old
Who wept when daddy went away.
Did I not take you in my heart and keep you there?

You think "Daddy is gone"
And sometimes a tear
Steals across your cheek
When you remember how we romped
Like two carefree children.

Together we played "Bear."
Screaming with delight.
Or told marvelous stories.
We walked through Formosan lanes,
And you prayed at my knee.

My little girl.
Sweet blond angel
With unwashed face,
Too careless with your clothes.
Alas, alas, my Charlotte.

Could I forget you?
Ah no, dear child.
I feel your little arms around my neck.
And hear your laughing voice.
Though half a world away.

The poem above captures part of my anguish of having a father who was often absent. Even now I wonder what he was doing in London then. The people who knew him and supported his ministry were in America. Had he gone to London to sell oriental antiques to keep his ministry going? Was he doing some type of research? (His book *The Asiatic Fathers of America* said that all of his life he was "searching, searching, searching" regarding whether ancient Chinese had traveled to America.)

Apparently he also felt the sting of the separation of miles, but did he really understand why I cried? Did he not realize that keeping me in his heart was not equivalent to having my arms around his neck?

At that point in time my Father had a wife and four children. My Mother had just found out that she was pregnant with a fifth child (John Clifford). Why was the poem written to me? Perhaps I was the one who had cried the loudest and clung to him when he left. My brother Hendon III, who would soon be three, perhaps was too young to realize what was going on when father left. Perhaps my sisters and mother had settled in to the fact that Dad traveled often. Regardless, his absences hurt all of us.

Despite Father's restlessness, the poem below shows his admission that he sometimes went with "trembling steps" to fulfill his calling in life.

"Song of the Night and the Day"

Laughter and Tears
By Hendon Harris Jr.
October 1953, Taipei

I have rejoiced in the night and the day.
Sorrow I've known, and joy, on the way.
Have drunk the rich wine of love's delight,
Sung the song of the day and hymn of the night.

I have experienced the night and the day.
Seen the freshening dawn and the sunset bay.
Have laughed with the careless and wept with the sad,
Admired the good and bemoaned the bad.

Courage, Endurance, Sacrifice

I have believed through the night and the day
That truth will live and falsehood decay
Since God abides through shadow and light
As King of the day and emperor of night.

So I've gone forth by night and by day
To duties at home or far away.
With trembling steps – yet still have gone –
To the battle field of the coming dawn.

And I have dreamed through the night and the day.
That our greed and sin will pass away.
And the Lord of glory descend in light
To renew the day, and abolish the night.

My father moved our family to Taiwan with the expectation that he would be able to continue preaching to the Chinese troops. However, as more missionaries began moving into Taiwan some of them asked why Harris was allowed to preach to the troops and not them. Therefore, the government soon ended that opportunity – much to Hendon Jr's disappointment. It hurt him deeply that it was the action of other missionaries that took away his opportunity to preach to the soldiers. The poem below was written about that time.

"I Heard Great Music"

Wings of the Morning
Hendon Harris Jr.
November, 1951

I heard great music,
And my soul leaped forth
As aircraft run to meet the sky,
In thundering effort at the take-off time,
When man must climb to heaven or die.

I heard great music,
And re-wove anew
The nail-torn fabric of my Soul
Forgot old hurts and bitterness;
Found a new gladness, and a grander goal.

I heard great music,
On an awful day,
When friendship's strongest bonds had failed
But God spoke firmly through the glowing notes
My strength returned and heaven prevailed

I head great music,
And I rose again,
From cruel prostration in the ruins of life
Turned once again to duty and to love,
To meet my hour and to face my strife.

I heard great music,
And the rushing tears
Washed off the mingled dust of pain and grief
I gathered roses from the barren snows,
And doubts dissolved before a fresh belief.

The Orphanage at Banqiao

As Father's grandparents, William and Florence Powell, had established in Mexico, my parents similarly ran an orphanage in the small village of Banqiao, Taiwan, from the beginning of our time there.

Our family resided in one end of the simple orphanage building which had cement floors. On our hard wooden beds, the same as for the orphans, were thin straw mats where we slept under mosquito nets. There was no running water – just a hand pump in the front yard. We could not drink the water unless it was boiled. All that Mom had to cook on was a wok over a charcoal fire pit. Bath water had to be heated over the same fire.

Mother could barely speak any Chinese when she made her first grocery shopping trip to the open air market where people bartered for food. Sometimes necessity makes one learn faster. Fortunately Mom was a good sport and had learned a few basics including numbers.

The meat in the market in Banqiao was covered with flies. That must have been difficult for Mom, a registered nurse, who was used to sanitation. The butcher would cut off whatever slab she pointed to then handed it to her unwrapped, hanging on a string. The flies followed as she carried it home.

Usually the meat was very tough because at that time the people there normally waited until the water buffalo was about ready to die before butchering it. Mom found creative ways to cook the meat so it would be sanitary and tender enough to chew. Her pressure cooker was a necessity. Since it was weeks before the trunks arrived which contained our eating utensils, we had to quickly learn to use chopsticks or starve.

Our milk, which Mom insisted that we drink, was made from boiled water mixed with milk powder. At that time none of the Chinese there drank milk or used dairy products. Despite the fact that Mom kept us clean, on more than one occasion Chinese complained to her that we smelled to them like sour milk. That criticism came while we were coping with odors from their open sewers.

Both the foods and table manners there were quite different from what we had been accustomed to in America. Father insisted that we adapt to Chinese culture. For breakfast instead of cornflakes, which we could not find there, we ate Chinese fare – thin watery rice containing peanuts. We called it "soupy rice." In America we had been taught not to slurp our food or to hold a bowl to our mouth. The custom there was to eat the soupy rice with chopsticks and with bowl held up to the mouth to slurp it down. Belching, which we had been taught not to do in America, was there considered a compliment to the chef.

We learned to politely observe chicken and fish served at restaurants with the heads on. None of us ever fought for the heads. Under Dad's insistence and watchful eye we reluctantly ate century eggs. Those eggs had been buried in the ground until the yolk was iridescent green. (There really is such a thing as green eggs and ham!)

We learned to love squid jerky, a very salty dried fruit, spring rolls, delightful tropical fruits, and many other types of Chinese delicacies. Now whenever our family is all together and we go out to eat, it is almost always Chinese food.

Currently Banqiao is a huge metropolis, but at that time it was rural. My sisters had brought clamp-on roller skates from America. The only place in town with an open concrete area suitable for skating was the Chinese military compound. The soldiers allowed us to skate there on their smooth pavement. Only five years old and just learning to skate, I fell frequently. The military men took great sport in laughing boisterously whenever I fell.

Although he had never skated before, the commanding officer decided that surely he could skate better than I was. Therefore, my sisters adjusted the skates and clamped them onto his shoes. On his first try the commander fell to the ground in front of his men. When he fell, all of the Harris siblings laughed uproariously, but his men were deathly silent.

Whenever there was an earthquake, and it seemed that happened frequently, an official from the town would come to our home to tell us to get out of the building for fear it would fall on us. Usually we were terrified. Once when there were several aftershocks and her bath had been repeatedly interrupted, Lillian stubbornly decided to just stay put in the tin tub.

As Caucasians we were an oddity in town and everyone seemed filled with curiosity. Mother was quite upset several times when she bathed us girls and saw Chinese men, who had scaled the fence, looking in the windows. Normally Mom was calm and adapted to almost everything in Taiwan, but "peeping Toms" were beyond her tolerance level. Each time she shrieked and chased them off.

The first Christmas we found a little tree of some variety. However, we had no Christmas ornaments with which to decorate it, and certainly there would be no store where we could find such treasures. We children had discovered in our forays a little factory that made tin cans. Discarded shavings from the tin cans formed dangerously sharp but shiny spirals. Ignoring the danger, that Christmas we used those beautiful spirals as decorations for our tree.

In our trips around the village we were always in search of candy. It seemed strange to us that no one in our village besides us had a sweet tooth. One day we found something that was wrapped in brightly colored paper. We were excitedly optimistic that we had found candy. However, after purchasing and opening it, we were disappointed that it was a pickled prune.

The children in our family soon learned that street vendors sold sticks of sugar cane which were keep moist by soaking in questionably sanitary water. Experiencing the sweet taste of sugar cane required a lot of chewing and spitting of cane fiber. Stories of our forays and new and exotic findings were shared with our interested parents at the dinner table each night.

Dressed in a ruffled smock uniform, like all the students in my class, I attended Chinese kindergarten in Banqiao. We learned delightful little songs in Mandarin including one about calling a pet kitten to eat. I was intrigued that Chinese and American kittens both say "meow." Therefore, I reasoned that American and Chinese cats both spoke the same language. No "monkey business" of learning a foreign language for them.

In a daily ritual at my school the kindergarten students bowed to a huge picture of Sun Yat-sen. At that tender age I knew nothing of politics. However, my parents thought that bowing to Sun was ancestor worship and thus not appropriate for an American Christian. They sought and received permission for me to abstain from that tradition.

My sisters Marjorie and Lillian ages eight and nine took a one hour train ride each way each weekday into Taipei to attend the American school there. However, on the long, crowded train rides they repeatedly contracted head lice. Each time they got lice we all had to be treated.

We children all also got intestinal worms so were de-wormed on a regular basis. De-worming was an unpleasant experience, but out of our parents sight we had contests to see which one of us would pass the longest worm.

Other maladies that different family members contracted in Taiwan included amoebic dysentery, carbuncles, and tuberculosis. Father repeatedly had malaria. Once mother suffered with boils all over her body. Young Hendon III repeatedly got boils on his head. Mother thought that it was from everyone within reach patting his blond hair.

One morning we woke up to find out that the previous evening our father had been in a jeep wreck while going out to an evening meeting. Fortunately his injuries were not permanent.

"What Music!"

Wings of the Morning
By Hendon M. Harris Jr.

What music doth the Christian know
In all his wanderings here below!
The trees and mountains sing an air
Of heavenly life beyond compare.
And in all anthems interwove,
The rich notes of his Savior's love.

What calmness doth the Christian find.
As moves the Spirit through his mind!
The holy Friend who quiets all fears,
And wipes away our anguished tears.
When He abides all conflicts cease
While soft we rest on wings of peace.

Courage, Endurance, Sacrifice

What sacred awe the Christian knows
In viewing Jesus; wounds and woes
For him, the crimson fountain spilt
For him, God's tears to purge his guilt.
And thankfully he bows to pray
Before the gift beyond repay.

What promises the Christian hears!
What future joys when Christ appears!
He views the earth with quiet eyes,
Like airmen ere they climb the skies.
Before his vibrant soul breaks free
And rises to eternity.

After a period of time it became apparent that some of the "orphans" in my parents orphanage were not really parentless but had been sent there by relatives hoping they would get benefits or education that the Americans might afford them. Later when we moved out of the village and into Taipei, my father left the orphanage in charge of a trusted Chinese man.

A few months after that my parents were told that the director of that institution was stealing funds intended for the orphans – leaving the children with inadequate provision. When my father had to replace that man with someone else, the first director became very irate. One of the orphans secretly got word to my parents that the former director was coming to visit us but planned while there to murder my father's eldest son (the ultimate Chinese revenge). The man did come, but my parents were very much on guard during his visit.

The poems below are from *Poems for Grown up Children,*[428] which my father wrote and dedicated "to my children, Marjorie Florence, Lillian, Charlotte, Hendon [III], John Clifford, and Aurora Dawn." (The youngest child, Mejchahl, was not yet on the scene). All of these were written in the early 1950's while we lived in Taiwan. At this late date it is wonderful to have these warm sentiments that were expressed directly from Father's heart.

"Journey into the Fields"

By Hendon Harris Jr.

"Come, Daddy!" cried the children
"Come and see all of our forts."
So I followed down the path way
Searching out their youthful sports.

I left time along the road way.
And became a boy again,
Singing through the fields of childhood,
Leaping through each glorious glen.

And once more delightful mystery
Beckoned from the hills ahead
"There's the robber's cave" cried Lillian
"Here's the place the goblins fled."

"Lion Rock" above the river
Filled the heart with pleasant fear.
We must walk with cautious footsteps,
"Over there" were feeding deer.

"Look, the hidden slide!" called Charlotte
Then she slid, to prove its worth.
And I chased my little pirates
Over rich delightful earth.

But when the sun was sinking
We went back to that gray world
Where the sweetest songs are muffled
And the gayest flags are furled.

The poem above reminds me that when my father was a child, for safety reasons, he was kept within a compound. He seldom if ever as a young boy in China had the delights that we experienced together in Taiwan.

"The Unwashed Boy"

By Hendon Harris Jr.

When I sit at dinner table
Daddy cries, "Look at those hands!
That boy really is unable,
or perhaps misunderstands.

"Doesn't he know soap or use it?
Have you even seen him wash?
We've a scrub brush, did he lose it?
Does he think I'm talking bosh?"

Poor dad is just too prissy
And his standards much too high
Does he want me be a sissy,
Or a useful half-washed guy?

Charlotte on water buffalo in Taiwan ca 1954. Brother John is the blond child in the foreground.

"Gather Flowers"

By Hendon Harris Jr.

Gather flowers while you may
Childhood and youth soon pass away
First the spring and next the prime
Then the fall and winter time

Gather blossoms while you can
For life is swift
My little man.

Aurora Dawn, our parent's sixth child (now an attorney in Southern California) arrived a year after our brother John. We still joke that both of them were "Made in Taiwan."

"Our Baby"

By Hendon Harris Jr.

Our baby's called Aurora Dawn
And she's sweet enough to eat.
She has the cutest little hands
And two rosy little feet.

Charlotte Harris Rees

She smiles and gurgles when you laugh
And cries if you seem to frown
She has no hair but she's pretty fair
When dressed her long white gown

She looks like a wise old lady
When her bonnet's on her head,
But when she hollers in the night
It's enough to wake the dead.

I can hardly wait until she's big
But mama won't agree.
She says they grow up much too fast
And that's what I can't see.

"When at Night"

By Hendon Harris Jr.

When at night I kneel to pray
And see the sky-lamps far away
I wonder, as I watch a star,
How God can hear my prayers so far.

But when I see my mother's face
And all the good things at our place
I know that God my prayers can hear
Because His kindness is so near.

"Gentle Jesus Kind and Good"

By Hendon Harris Jr.

Gentle Jesus, kind and good
Lead me every day.
I have known too much of sin
Too little of Thy way.

Gentle Jesus, let me be.
Like a child once more.
Guide me to eternal life
And to Thy native shore.

Gentle Jesus, in Your hand
Cruel wounds I see.
Savior, do I understand
It was marred for me?

Courage, Endurance, Sacrifice

Gentle Jesus, grant me this,
Never leave my side.
Take me to that land of bliss
Where Your white lambs abide.

Father's book of poems entitled *Laughter and Tears*[429] was dedicated "to my wife Marjorie." The following are from that book:

"My Gentle Love"

By Hendon Harris Jr.

My gentle love for you
Is like that quiet wind
Which warmly blows in April
Promising soft rains to send.

I love you mildly as those stars
Which twinkle kindly in the summer blue
Unmoved by any passing storms
Unto the heavens ever true

But still, my burning love
Is wild as those cruel waves which fly
Shoreward, with angry raging,
Beating the rocks with anguished cry.

"The Gift"

By Hendon Harris Jr.

I have no silken scrolls of rarest hue,
On which to mark fine words of tact and charm;
No silver plaques can I present to you,
Engraved with phrases which disarm.

Yet read, O princess of the golden age,
How through the twilight of time's fading street,
I come with love's rich equipage,
To lay my heart before your feet.

"My Desire"

By Hendon Harris Jr.

Let me have a true friend,
And a few honest tears
For the sadness of earth.
Let me learn to love my God.

May my children, coming after,
Know my Savior and my song
As they sing along
The road of life.

Let me take up arms
To aid the weak
And be courageous
And clean.

Let me climb God's ladder
And swing like a child
Over the sunny world
Drinking spiritual mead.

And when I die,
Let me plunge like a meteor,
Into Love's sea;
Into God's arms,
And immortality.

The Mountains of Taiwan

Instead of just doing church work in the easier areas, my parents also chose to minister to the aboriginal mountain people of Taiwan. That involved long trips while hiking up the mountains. I remember crossing a swaying suspension foot bridge over a deep gorge then riding down the mountain at break neck speed on a narrow gauge car with a simple hand brake.

Even at that young age I realized that the mountain people were very different from the people in the low lands. They lived a primitive life. Once when a group of aborigines came to our home (then in Taipei), Mom noticed that one of the men stayed too long in the bathroom. On investigating afterwards she discovered that every one of our toothbrushes was wet. Apparently he had tried them all!

Hendon Jr. with Chinese evangelist and mountain man and his monkey.

Some mountain people in their native costumes.

However, we loved the sweet mountain people. One of our amahs came from them. The translation of her name was "Beautiful Banana."

One group of aborigines visited us on a rainy day. Mom had placed a mat in front of the door so that they could wipe their feet. Every one of them gingerly stepped around the door mat – not wanting to dirty it.

Father's poem below depicts one trip down the mountain when a low hanging electrical wire narrowly missed beheading Father. The village chief severely cut his hand in reaching out to lift up the wire. Those recent head hunters had been animists, terrified by many things. When they turned to Christianity they burned their good luck charms. They delighted in my father's visits.

"The Mad Little River"

(The Fu Shan River above Wulai, Formosa)

By Hendon Harris Jr.

It's only a river, a mad little river
North of Wulai, and the push car railway
Ran along and above it, a thousand feet above it,
On the afternoon we climbed up the heights
To the mountain village.

But the little river, the mad little river
Caught a bit of my heart that I can't retrieve.
The Atayal [aboriginal] girls danced their praise to God
In the bamboo church, and the chief leaped for joy
And they threw me a bouquet of friendship
In the little village above the mad river.

The chief and I cleaned stones from the river
From the frothing and plunging determined current
Cold and clear on that rainy winter day.
And we baptized them near some sheltering rocks
First the chief, then twenty-five of his people
With tattoo marks of their tribal wars.

We slept on mats, three Chinese and I
But we could hear the crying of the night river
We were not forsaken, for there was God's voice
'Twas the roar of hope, comforting frenzy,
Overflowing splendor, prize song of rain
Singing down the valleys the rich aria of life.

'Tis only a simple forthright story
There were landslides on the way back
And the chief cut his hand saving us
When a low wire nearly beheaded us
As our little car raced down the mountain
Above that frantic torrent of delight.

They promise a river in heaven
To ease the pain of hurt hearts
And I hope that little river is allowed to flow there
Resting after its raging and hopes.
For I'm sure it had a motive and dream
A driving, insatiable, beautiful purpose
Making it sing and leap that day toward the golden Taiwan Sea.

Hendon Harris Jr.

"The Mountain Path"

By Hendon Harris Jr.
Formosa, Feb., 1954

We traveled fast around the cliffs
And far below the shimmering waters
Roared in delirious joy
The dazzling heights increase
And the slippery way contracted.

A wave of dizziness struck the leader
One slip meant death. How plain the ending.
"Body never recovered, fell direct to stream."
He shook his head – mind must be clear.
"One step, one step, enough for me."

A pile of gravel, in the path,
And if it slid? – O well.
He placed his foot upon it and it held
Long enough for him to reach the rock.

"If you lost one of us here
Many would blame you." It was Wang,
[probably a reference to his co-worker Wang Shi Ping]
Not critical, but simply stating facts.

"Were it not for human souls
I would never lead you here."
The Chinese nodded, satisfied,
And followed firmly on that lofty way.

In Taiwan my parents, with the help of Chinese believers whom they had trained, started twenty one chapels and churches, had a Bible training school, an orphanage, and a work among the mountain people. As well as reaching out to help save men's souls, on more than one occasion through a heroic act Father saved a man's physical life as well.

One of the bamboo churches in Taiwan.

It is my understanding that the believers in the mountains of Taiwan thrive in their faith even today.

Father's faithful friend and co-worker, Wen Yeh Ching, continued the work for many years after we moved away. He had become a Christian in 1949. A college graduate, he also studied at the Hundred Nations Crusade Seminary for four years before beginning work in the mountains.

One of the mountain pastors was Cheng Lin Tsao. He was of the Paiwan tribe and had graduated in 1955 from the Hundred Nations Bible School. He was ordained in 1968.

Father visited periodically bringing financial support and encouragement. Churches located there included Feng-Li, Li-Chuo, Cheih-Ta, Chia-Lan, Chieh-Shing, Shih-man, Lai-1, Nan-Heh, and Ku-lou. There were also congregations in the villages of Fong Yeh, Ta Ma Li, Tu Li, Pai Shou Lin, Tai Pong, Tu Pong, Ta Shi, Shin Yuan (Tai Tung county), I Ling, Nang Ho, Kuei Tsung, and Chi Chia (Ping Tung).[430]

Che-ji-do (Jejudo)

Several times in his life Father felt what he believed were strong callings from God to take action. One had come in 1950 telling him to go to Taiwan. In 1953, at the end of the Korean War, a call came again. Chinese prisoners of war (POW) who had been fighting on the side of the North Koreans but had been captured during that conflict had to decide whether they wanted to return to China via North Korea or go to Taiwan to live. According to one news report approximately six to seven thousand of those POWs were but 13 or 14 years old.

Father wrote in *The Asiatic Fathers of America*:

> In 1953 when the Americans had 14,500 anti-communist prisoners of war in Korea, the writer [Hendon Jr.] felt that God wanted him to help the Chinese prisoners. He flew to Korea, though he was sick at the time, and stayed in a missionary's house for two weeks, praying for the Chinese prisoners.
>
> At the end of that time Colonel Hanson flew him down to Che-ji-do Island, where the POWs were very reluctant to go up to Pan Mun Jom to be interviewed by the Reds [Chinese Communists]. But he spoke to these prisoners 31 times – until all the skin peeled off his face in the hot sun [from sun burn] and told them to trust in God and to resist the North Korean overtures.... and promised that the Americans would convey them to Taiwan.
>
> They believed his words, signified their willingness to trust Christ...and he finally saw most of them alive and well on the island of Formosa. Is it possible that his prayers and faith in God played a small part in bringing them to spiritual and physical liberty and happiness?[431]

An Associated Press article (shown below) appeared in Taipei's *China Post* on August 12, 1953.

A Pastor's Story Of Chinese POWs

Taipei, Aug. 10 (AP) A China-born American Baptist missionary is back in Taipei after having spoken to all but 1,000 of the anti-Communist Chinese war prisoners in Korea against yielding to persuasion to return to Red China.

Dr. Herndon Harris of Chicago, Illinois, said the prisoners seemed resolved not to go back, but he feared some of the weaker ones might yield to pressure.

Harris, former pastor at the Airport Baptist Church in Chicago, said he made 31 addresses of an hour each to Chinese war prisoners in difference enclosures.

Altogether, 13,500 heard him.

Harris originally came to Formosa three years ago after seeing a vision which he interpreted as a divine call for him to help Chiang Kai-shek's Nationalists.

His visit to Korea, he said, was also impelled by God who, he felt, wanted him to help in preparing the anti-Communist Chinese prisoners for their "coming ordeal at the hands of Communist inquisitors."

In this addresses to the presoners, Harris assured them that they would be given a warm welcome in Formosa.

They had their choice, he told them, of eating the fruits of two trees, one of freedom: in Formosa, or the other, of slavery, in Red China.

He asked them to place their faith in God, and God would help them.

Harris also sang to the prisoners a song he had written, "the song of Taiwan" (Taiwan, which means "terraced bay," is the Chinese name for Formosa, Portuguese for "isle beautiful.")

The prisoners learned the song and sang it themselves.

Every single one of the prisoners, he asserted, bore the tattooed inscription "oppose Communism resist Russia."

In the compounds, he said, he saw portraits of President Chiang Kai-shek, Nationalist flags and maps of Formosa.

Harris believed he might have helped to allay some of the unsureness the prisoners felt over their future.

Great credit, he said, must be given to the C.I.E. and the Psychological Warfare Section of the U.N. Command for the work they had done.

Harris wife and five children joined him in Formosa two years ago.

Harris is sponsoring a plan to build low cost bamboo churches in Formosa.

So far he has built three at a cost of US$150 apiece.

Apparently Father had been chosen to speak to the POWs because of his command of Mandarin, which he spoke fluently with the Kaifeng accent of his boyhood days. The Nationalist Chinese ambassador to Korea, Mr. Wang Don Yuan, warmly welcomed him.

At that time the newspapers of Korea and Taiwan also published information concerning Hendon Jr.'s trip to Korea. The English translation of one of those articles is shown at the end of this book.

Most of those Chinese POWs to whom Hendon Jr. spoke in Korea chose to go to Taiwan rather than return to China. Although Hendon Jr. was just an ordinary spectator in the crowd the day they arrived, the soldiers recognized him. The convoy stopped, they greeted him, and he warmly welcomed them. See the photos attached of Father with some of those soldiers on the day they arrived in Taiwan.

Dr. Hendon Harris Jr. welcoming Chinese POWs to Taiwan.

Upon arriving in Taiwan, the Chinese POWs were excited to see Dr. Harris. Many had prayed in Korea to receive Christ when Harris visited the POW camps there.

Later Father was commended by Vice President Richard Nixon and by the U.S. Ambassador to Taiwan, Karl Rankin, for his service with those prisoners of war.

Did the Vice - President,

Mr. Nixon, Commend the

Work of the

DURING MR. RICHARD NIXON'S recent tour of the Orient, Mr. Karl Rankin, our very capable American ambassador to China, was so impressed by God's marvelous blessing on our Society's work in preparing the Chinese on Chejido for their experiences at Pan Mun Jom — that he invited the Director, Dr. Hendon Harris, to meet the Vice-president of the United States, who thanked us for our good work in Korea.

Baptist

EVANGELIZATION
SOCIETY
IN THE FAR EAST?

Central Office:
FIRST BAPTIST CHURCH
First Street at Pennsylvania
Hobart, Indiana

A column was published in which American Vice President Richard Nixon highly praised Dr. Hendon Harris Jr. for the fruit of his godly labors to the prisoner of war camp in Korea.

In a signed statement dated November 9, 1955 Chief Zhang Han Guang, of the Nationalist Chinese President's Palace in Taiwan bore witness to Dr. Harris's accomplishments, photos, and the letters and other items in the Harris album.[432]

Taiwan and South Korea still celebrate World Freedom Day in January of each year in honor of those over 14,000 soldiers who arrived in Taiwan on January 23, 1954.[433] The report is that those soldiers were warmly welcomed by the Chinese in Taiwan and each given an allotment of land there and that some eventually became quite wealthy.

Four of the Harris Children Leave Taiwan

For years often air raid warning sirens blared at night in Taiwan as bright search lights scanned the sky looking for planes. My parents, like many others of that era, had an air raid shelter built in our front yard. In 1955, because of the continuing threat, our parents sent their four oldest children to a boarding school in the US. They feared that if they had to leave Taiwan in a hurry that it would be very difficult to do so with all six of us children.

Hendon Jr. had grown up with missionary children being sent away to boarding school as a rite of passage. However, being separated from her children was a bitter pill for our mother to swallow. Probably making it even more difficult for my mother was the fact that when my sisters, my brother, and I left for America I cried loudly and had to be pushed onto the airplane.

Hendon Harris Jr. family on the day in 1955 the four oldest Harris children left to go to boarding school.

On the three day trip by plane across the Pacific and on to Chicago our 13 year old sister, Marjorie, with our travel documents safety pinned in her bra, was in charge of us – except for the one day we spent in Hawaii with a missionary family. Our prop plane refueled in Manilla, Guam, Wake, Midway, Honolulu, and Los Angeles before going on to Chicago.

Lillian startled US Customs and Immigrations officials in Hawaii when they saw her horn collection. (They feared that she was bringing in hoof and horn disease.) Her horns were confiscated and our suitcases carefully searched for other possible contraband.

In Hawaii I finally found sweets – then ate too many. Back on the plane during a bumpy flight I threw up all over the uniform of a flight attendant. During that same storm just when Marjorie tried to stand, the plane jerked. She reached for the back of the seat in front of her but instead grabbed the head of the man in that seat.

Father had selected French Camp Academy in Mississippi, not far from where his parents lived, as our boarding school. However, our grandparents seldom visited. Even when we went to their home for Christmas they woke us up early on Christmas morning, gave us presents, then sent us back to the boarding school by early Christmas afternoon.

During that visit I was so sick with what must have been strep throat that I could not eat any of the Christmas goodies that Grandmother Harris prepared. However, no one took me to a doctor. It was a sad lonely time for all of us.

Once I tried running away with some other girls, but there was nowhere to run. Within a few hours we were back at the boarding school. Fortunately at the day school associated with French Camp Academy my fourth grade teacher took a special interest in me. She was a godsend during that lonely school year.

Hendon III, only six years old, had the hardest time of any of us Harris siblings with the adjustment of being away from family. While being driven down to Mississippi from Chicago by father's pastor friend, we had to stop at a hospital to have a huge boil on Hendon III's head lanced. The hospital wrapped Hendon's head in a turban of bandages. When he arrived at the school so bedecked, and the children learned that his name was Hendon, they decided to call him "Hindu" instead.

Boys and girls at that school were separated, so Hendon had little contact with his sisters. Perhaps from the trauma, Hendon soon forgot his Mandarin, which at one time he spoke flawlessly.

Hendon Jr. and Marjorie lived in Taiwan until 1956 when they relocated back to the United States. Our mother could not stand the separation from her children and by the end of that school year in spring 1956 she arrived at French Camp with the two youngest children (John and Aurora Dawn) and our pet monkey.

Shortly after the Harris siblings were reunited.

Life in Indiana

The family then settled in rural Worthington, Indiana. There the cost of living was low, a hospital where mother could work was not far away, and we were in the middle of the continent – affording Father the ability to easily travel either east or west. One year he put over 100 thousand miles on a brand new car as he continued raising funds for the ministries in Taiwan.

Before we moved to Indiana Mother had never learned to drive, but that was the only way for her to get to work. Therefore, sometimes with six children in the car she taught herself to drive a stick shift automobile as we rocked and rolled down those rural roads.

Dr. Hendon Harris Jr. leaving on another trip.

Periodically Father would go back to Taiwan to check on the work there. An article titled "A Friend of China" (shown in addendums at back of this book) was published during one of those trips back to Taiwan. That article mentioned that Hendon Jr.'s movie "The Challenge of Formosa" was being shown in the US and Canada.

THE
CHALLENGE
OF
FORMOSA

a forty-five minute
TECHNICOLOR MOVIE

Presented by

The

Baptist Evangelization Society

Produced by

Hendon and Marjorie Harris

filmed in
FORMOSA

Once my father brought over to America his faithful co-worker, the former Chinese General, Wang Shi Ping, to speak in churches while Father translated.

Mr. Wang visited our home in Indiana. An excellent cook, he asked permission to prepare us a Chinese meal. We lived in the country and had large fresh eggs. On cracking open the first egg Mr. Wang was amazed that it had a double yolk. It was the only double yolk egg we had seen, but Wang immediately asked, "Do all eggs in America have double yolks?"

On one of their trips Wang was asleep in the back seat when Father stopped for gasoline. When Father went into the station to pay, Wang woke up and went to the restroom. Unaware that Wang was not still asleep in the back seat, Father drove on. Several miles down the road he was chased by a highway patrol car with blaring siren. Wang, who spoke almost no English, was returned to Father.

As had happened in Banqiao, in rural Indiana our family members were again perceived as oddities. In that era few people in that locality had even been out of Indiana, much less out of the United States. The locals regarded us as aliens. We thought of Chinese as friends so did not understand why some of the other children called us "Chinese" in an attempt to mock us. Yet at the same time we comprehended their tone of voice.

Three-year-old John shocked the congregation at a small rural church on the first Sunday we visited. After the service, before Mom could stop him, he urinated on the front steps of the chapel as everyone exited. (In Taiwan it was normal behavior for a small child to relieve himself on a public sidewalk.) When we were occasionally overheard speaking Mandarin to each other, to them it sounded like the babbling of lunatics.

It was while we were living in Indiana that it was discovered our sister Lillian had tuberculosis (TB). Apparently she had picked it up more than a year prior in Taiwan where it had been wide spread. She had to stay in a sanatorium a couple hours drive away during her freshman year of high school while she recuperated. The whole Worthington school system had to be medically tested for TB, so our family fell even further on the popularity scale.

As the only young person at the sanatorium, Lillian was lonely. The patients were not allowed to go outside. One resident told Lillian that she would never exit alive. Mom was only able to visit Lillian for an hour or two on Sunday afternoons while we children played on the lawn of that facility. We were only allowed to wave to Lillian from a distance. Fortunately within a year her TB was arrested, and she has lived a full and useful life since.

At that time the Indiana county where our family resided had a law stating that people of African American ancestry could not spend the night there. We lived near a highway. Once the car of an African American family traveling through had mechanical problems. Defying that law, my father brought them home to stay with us until their car could be repaired.

Father and mother started Morton Street Baptist Church in Bloomington, Indiana. Our family moved frequently as they continued "flipping houses" in various towns to help support the family and the ministry. In five years' time we lived first in Worthington, then two different houses in Bloomington, Winona Lake, and Wabash – all in Indiana. Each time we children had to re-establish ourselves in new schools.

Sometime during the period 1957-1960 Hendon Jr. spoke at Columbia Bible College. Just as his grandfather, William Powell's inspiring chapel message had been remembered by a college student many years later, so was Hendon Jr.'s.

Dr. Bob Alderman was for many years the Head Pastor at Shenandoah Baptist Church in Roanoke, Virginia, but now is Minister at Large. In 2016 he wrote:

> During my days (1957-1960) as a part-time student at Columbia Bible College…I had the opportunity to hear Missionary Hendon Harris Jr. as one of our chapel speakers. His message challenged me in such a way that even now, more than 57 years later, I not only remember his name (*it is an unusual thing for a student to remember the names of chapel speakers*), but I also remember the essence of his message. It was the essence of his message that imbedded itself in my mind, and accordingly prompted me to remember his name through the years.
>
> The first essence was his evident conviction and devotion to the call of world missions. There seemed to be no doubt in my mind that Dr. Harris was deeply schooled in and devoted to what we now refer to as the "Great Commission" of our Lord and Savior Jesus Christ.
>
> There was yet another factor in the impact of his message on my life. It was not only that he had a devotion to his calling, but that he also had a strategy of responding to and fulfilling that calling.
>
> ….such a vision and such a plan was very impressive to this young student. There seemed to be nothing small about his devotion and his goal. That was the kind of challenge with which I had been raised and accordingly which I still remember and sincerely appreciate. I am indebted to all who have come before us who call us to focus with truthful teaching and strategic planning on the matters of eternal worth.[434]

Subsequently, Dr. Alderman's church for many years has generously financially supported numerous missionary ministries.

My Grandmother Harris told that my father preached his first sermon at a very young age when he spoke about the sun, moon, and stars. I have no reason to doubt her. Father always had the gift of gab. Public speaking came naturally to him. In addition to his preaching ministry my father continued to spend much time raising money for the mission work – requiring him to travel frequently.

Father inserted jokes into his long sermons, but church people in America later told me that they preferred to hear Mom speak. Mom spoke of her love of God as she told down to earth stories about life in a foreign country. She had a warm and caring spirit.

The Harris Family Returns to Asia

After a few years back in the States, Hendon Jr. and Marjorie decided to move the family back to Asia, but this time they expanded their work to the British Colony of Hong Kong.

"Song of Hong Kong"

Words and music by Hendon Harris Jr.

I'd like to sing you a song
About the green hills of Hong Kong
A city so lovely and gay
Above the bright beautiful bay

Oh come, come, come along
To the beautiful tropical Isle of Hong Kong
Where they sing all the night
And work all the day
In the city above the blue bay, bay, bay.

At night every house and tong [district]
Shines bright on the hills of Hong Kong
Ten thousand jewels attire
A fairy gown sparkling with fire, fire, fire.

Hendon Jr. and Marjorie moved their family to Hong Kong in 1962. Daughters Lillian and Marjorie were in college by then so were left in the States. I was entering my junior year in high school. They wanted me to stay in the States, but I persuaded them to take me to Hong Kong, too. Over time several other Americans also ministered in my parents' organization alongside the Chinese workers.

When my family entered Hong Kong, my uncle, Lawrence Harris, who worked at the U.S. Embassy there, was leaving that colony on furlough. We arrived just in time for an exit dinner in his honor. That is the first time that I can remember meeting Uncle Lawrence and his family, my first cousins.

It had been arranged between the brothers in advance that we would stay in Lawrence's home during his year of absence. This afforded us a very nice place to stay in Kowloon.

Meeting the cousins- going clockwise from top center Charlotte, Larry, Hendon III, Ronald (barely visible), Jon, John (back to camera), David, Judith (twin of Ronald), Susan (twin of David), and Aurora Dawn. The other two children are unknown. All the above named are offspring of Hendon Jr. or Lawrence. Two of Hendon Jr.'s children (Marjorie and Lillian) were back in the States then and thus not in the photo.

At top of photo left to right Marguerite Harris, Lawrence Harris, Marjorie Harris, Hendon Harris Jr.

The seventh child in Hendon Jr.'s family, Mejchahl, was later adopted in Hong Kong. Mejchahl was one of my students when I was teaching English in a Chinese kindergarten. Mejchahl's English name was formed by Father by taking a letter from each family member's name – **M**arjorie, **E**lizabeth (for my mother – Marjorie Elizabeth), **J**ohn, **C**harlotte, **H**endon Jr., **A**urora, **H**endon III, and **L**illian.

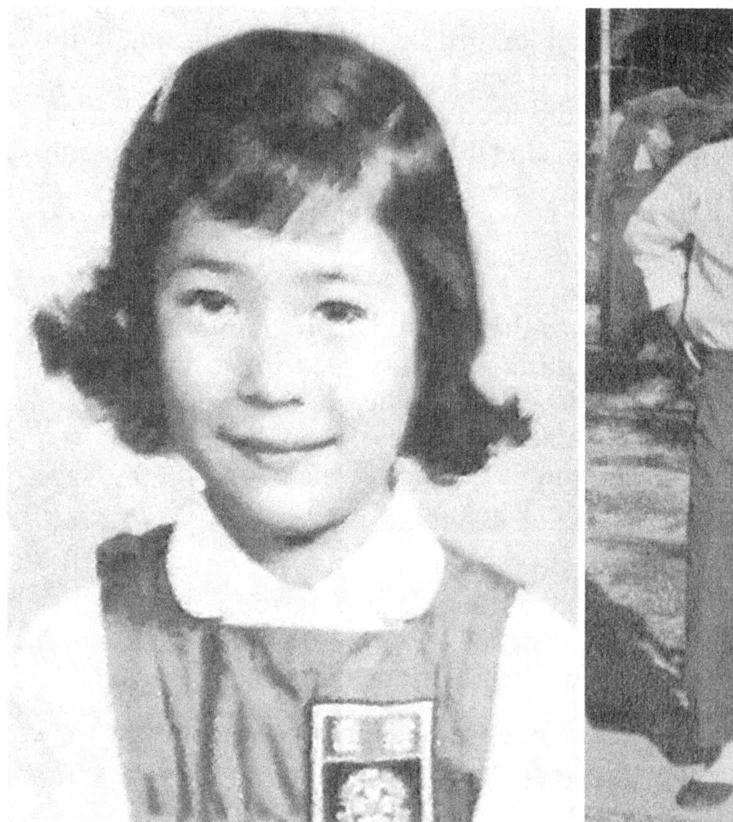

Our sister Mejchahl.

Shortly after we arrived in Hong Kong in October 1962 my parents invited a group to our residence for dinner. When they arrived two hours late another man showed up at the same time to see Father. My mother, thinking that he was one of the invited guests, seated him at the table, also. The man stayed throughout dinner. Unsure of American customs of hospitality, it was only after dinner that he revealed why he had come.

The story below dated December 3, 1962 is taken from my Hong Kong diary. I was teaching English in a Chinese kindergarten in the mornings and finishing my high school classes by correspondence in the afternoons.

> This noon when I got home from school a young man who had recently arrived from China was there. Mom had just come back from the immigration office, hadn't fixed lunch and was talking to the boy about God, so I fixed … soup and sandwiches. [My brother] John wasn't there yet

> so we left a bowl of soup at the end of the table for him. Hendon III [age 13] finished eating. While the rest of us were still at the table, he repeatedly tossed a ball in the air. Suddenly the ball slipped and hit dead center in the bowl of soup.
>
> Our guest was drenched with bright red tomato soup. The walls, the drapes, lamp shades, floor, table, people, and everything within a fifteen foot radius wore drops of tomato soup. We all jumped up and started cleaning – the guest first – of course. Hendon [III] was embarrassed beyond words and was afraid the guest thought he did it on purpose. We all laughed at the ridiculous mess. We apologized again and again to the guest, but he only laughed.

Another excerpt from my Hong Kong diary:

> December 10, 1962 - Tonight was the Christmas program at the school where I teach conversational English. I was assigned to sit on the front row to help entertain some important guests. Before the program started an American doctor who has studied Cantonese a month and can't even converse yet, read a speech in Cantonese. Some of the people on the stage had a hard time trying to keep from laughing.
>
> [In the Christmas program] The kindergarten had a nativity scene. The angels were up on a platform. One of them leaned on the stable, lost her balance, and while crying fell into the stable. As soon as she was taken off the stage, the wise men came up and nearly tripped over each other's robes.
>
> After the program the teachers and guests went out to a dinner. By the time we got back across the bay [Hong Kong Harbor] and onto the bus it was twelve-thirty a.m. As I walked home from the bus stop the streets were all empty and the street lights made me have about four shadows. My clicking heels echoed in the eerie silence, so I took them off and walked the rest of the distance from the bus stop home in my stocking feet.

(Even though in Hong Kong I sometimes returned home from meetings alone late at night, I don't recall my parents ever expressing safety concerns.)

> Dec. 14 - While Mom and I were shopping we passed a little woman selling milk tins from a box top. On the box from which she had taken the tins was written [in English]: "This is a donation from the people of the United States of America. It is to be distributed to the poor and not to be sold." Maybe that sign should have been…in Chinese.

In Hong Kong my parents worked in a church. The British school system included Bible classes in their curriculum so Mother taught Bible there. By that point my father had recruited several other American workers who had assignments of teaching in schools or evangelism.

Mom teaching in Hong Kong.

After a few months' time Uncle Lawrence decided not to return to his post in Hong Kong so gave up his residence there. We then moved to a home overlooking a Buddhist temple garden high on a mountainside in Shatin, New Territories. It was about an hour's train ride from Shatin to Kowloon where Westerners often went to "hang out" at the Y by Hong Kong harbor.

The only way to reach our new home in Shatin was to climb about 100 stair steps up the mountain – winding past small residences, businesses, and tea houses.

Although we had almost no yard – only a small terrace – our home was spacious with tiled floors and a great view of the harbor. My parents were able to rent the building for a good price because the Chinese considered it to be haunted.

View of Shatin Harbor from our home. The harbor has since been filled in.

My younger siblings attended King George V School in Hong Kong which utilized the British system through 11 years of schooling. However, I was in high school and needed a diploma from an American high school (the culmination of 12 years of schooling) to enter a university in the States.

While living in Shatin, I completed correspondence classes for American high school graduation as "Jack be nimble, Jack be quick, Jack go under the limbo stick" played loudly in English what seemed like all day every day on the Buddhist temple garden jukebox in the area below us. The click, click of mahjong tiles echoed night and day through the building next to us.

Father came home late one night. Walking the many steps up to our home he came across a profusely bleeding neighbor – a British businessman. As the inebriated man ascended the mountain he had fallen; his liquor bottle had broken; and he cut himself. With Father's help the man's life was spared. Later in appreciation the man gave Father an expensive chess set.

Below is an essay I wrote about my father in 1963.

My Dad Is Tops

> Preacher, poet, author, musician, actor, patriot, and business man – these are all words that describe my father. Although Daddy is only five feet and ten inches tall, most people think that he is much taller for he walks, talks, and thinks like a big man.
>
> Father's blond hair is almost gray now, but he still has a youthful spirit. He will accept anyone's challenge to a wrestling match, and he usually wins. In whatever game he plays or work he does, Father… plays to win. His strong feelings for justice and clean living sometimes seem fanatical to some people, but they show me that my dad is not an average man. He would not be afraid to stand up for right even if the whole world were against him.
>
> I am not so blind as to say that my father has no faults…. What I do say is that his good points outweigh his shortcomings by far. I am proud of him. I have never met another dad who could begin to compare to mine, and that is why I say that my dad is tops.

Both of my parents were very creative people. Many times Father came up with ideas and Mother would implement them. Mother's willingness to adapt to every change and her loyal devotion to Father and us was what held our family together.

In the early years it was easy to be close to Father. He did lots of things with his children then and told us wonderful stories. However, as we got older and especially after we married, it was difficult for him to relate to us. As we became adults, we all wanted to make the transition from the role of child to that of a friend to our parents. Most of us never got there with our father. For some reason he kept us at a distance. However, Mom always was our friend. I'm sure that was not easy for her – especially during our teen years.

When we arrived in Hong Kong Hendon III was 13 and I was 16. We were both finding our way in a new culture. Another excerpt from my diary states:

> Dec. 16, 1962 - This morning we went to the Emmanuel Church. My brother Hendon was mad because he wanted to go to the Southern Baptist church (where he had met some friends). When we got inside they were praying so we waited in the back. They prayed about five minutes and we saw no seats except in the very front, so Hendon left and went to the other church. I was mad because Hendon left, and (my younger brother and sister) Johnnie (11) and Dawn (10) embarrassed me by not being quiet during prayer – so I left. Mom told me to come back, but I didn't want to. We went home. On the way we had a heart to heart talk and I had a good cry. Before we reached home everything was settled.

Now as I read this I see Mom's calm self-control in a difficult situation. Just days before she had received news that her mother had died. She was unable to return to the States for the funeral and was still mourning that loss. However, her love for us over rode the moment when anger with disobedient teens could have flared.

The letter below was written when I was 17 after I had flown back to the States from Hong Kong to attend university. After I left Hong Kong I was never able to even visit my parents' home again until after I had married and they were living in the States. I include the letter here because it shows a missionary's child's closeness to her parents but also the agony at separation.

> Dear Mother,
>
> Last night as I lay down to sleep, thoughts of our last months together flashed through my mind. Many pictures danced before me.
>
> I re-lived the day of the big rain. Once again we stood together on the front terrace, watching as great rivers of yellow water cascaded down the steps into the temple garden farther down the hill. Our umbrella was small, so we stood very close. The warmth of your body was comforting as we quietly stood watching. I felt like a little girl, and wished to never grow up.
>
> A new picture came. We were sitting together in the playroom. Pins, needles, patterns, thread, and all types of sewing were spread about. We had worked hard that day as we prepared my clothes for college. This was only one of many such days; yet you worked on, never complaining. Your poor, tired eyes could no longer see to thread the needle, but still you worked. [She needed new glasses so I threaded the needle many times.]
>
> We talked of many things, but we did not like to discuss my departure. When we did, it all seemed like a bad dream. Never have I had such a close friend as you were then. It was difficult to think of our being torn from each other.
>
> I remembered the days we shopped together, and a new picture came. We stood outside the window of an Indian shop. I admired a beautifully embroidered sweater. Although our finances were already low, you insisted on pricing it. You could not afford it; yet you wanted to give me all you possibly could.
>
> Finally the last week came. Together we went to buy my plane ticket. Together we bought my last few souvenirs and gifts for friends in America. Together we packed my suitcases.
>
> The nights were the most wonderful though. I slept in your big double bed with you during that last week. [Apparently Father was traveling]. Each night we talked into the wee hours. Neither of us could sleep. There was so much to say and so little time to say it in. We lay in the dark, grasping

hands. Often I choked back the large lump in my throat. Too soon, too soon I would be gone for good.

The last day came – the last look at our home, our pets, and our yard. The family hurried down the hill to catch the train. After the train ride was the taxi ride to the airport. In the crowded airport we all waited for the plane. There was not much to say any more. We just waited.

The flight number was called. We collected my cabin baggage, and you arranged it so that I could carry it easily. Down the line I went, kissing each member of the family goodbye. I fought to keep back the tears. At last I grabbed you. I trembled. I tried to control myself, but the tears gushed down.

After a minute I was calmed by your gentle voice. Determined to be brave, I went through the gate. Straight ahead I walked. Suddenly I turned. I must look just once more – to see your beautiful face.

There you stood by the fence. I ran to your side. I clutched you. We cried. Oh, if only I could stay with you forever – but, no, I must leave. I must grow up. This is life.

I grabbed my packages. Briskly I walked away.

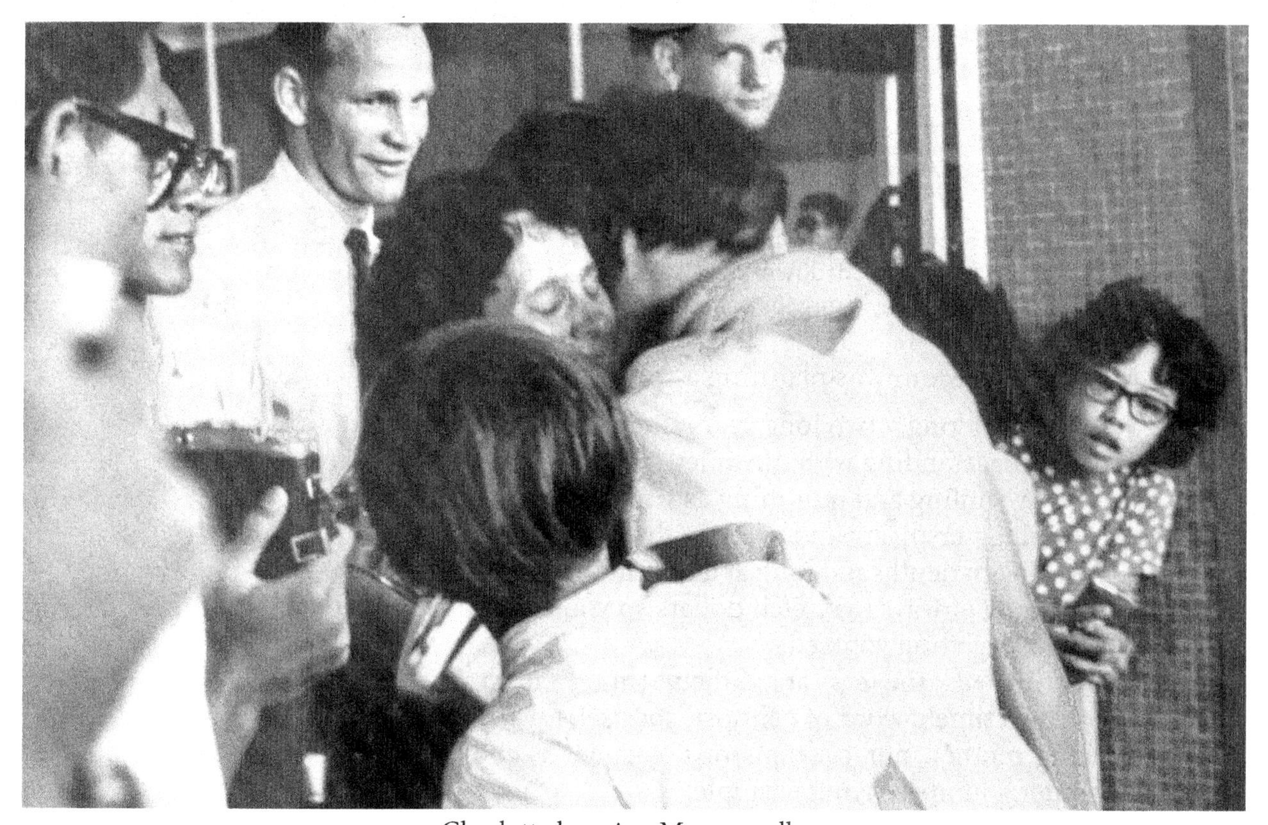

Charlotte hugging Mom goodbye.

Charlotte laden down with packages on Kai Tac Airport tarmac leaving Hong Kong for college.

Then in May 1965 as Mom was planning a trip back to the States I wrote:

> It has been long – two long years since last I saw you—
> Laughing, standing with me quietly in the rain
> Sweetly smiling and humming as you worked around the house
> Climbing perspiring and tired up the hill after a busy day in town
> Working patiently and long at a mountain of typing or sewing
> Scheming of ways to stretch dollars so you could buy more for us children
> While forgetting yourself,
> Fighting back the tears at a difficult time
> Giving a timely word of comfort, counsel, love, or rebuke,
> Being mother, nurse, seamstress, carpenter, cook, friend, teacher, typist, painter, plumber, living example,
> It has been long—two long years since last I saw you.

According to their letters at the time, the last full year (1966) that the Harris family lived in Hong Kong they had seen 302 people accept the Lord in their ministry which included a church, school chapel services, weekly Bible club, as well as 26 public school Bible classes per week. Two student missionaries from America assisted them that year.[435]

My mother Marjorie suffered from various ailments almost the whole time she lived in Hong Kong. Before moving there she had been in an auto accident when another driver ran a stop sign and broadsided her vehicle. Her back was in pain for the rest of her life. In January 1967 she had surgery for a growth on her thyroid which proved benign but still left her weak. About that time my fiancé, not understanding diseases that aliens are susceptible to in foreign countries, asked me why there was ALWAYS someone sick in my family in Hong Kong.

Their April 1967 prayer letter stated that there had been around 1000 conversions during the four and a half years they had lived in Hong Kong.[436]

Starting early that same year there were Communist demonstrations in that British colony accompanied by considerable political unrest. In May 1967 Marjorie, Aurora, and Mejchahl, while out together one day, were assaulted on the street.

When they accidentally happened upon an anti-foreign demonstration, they had their clothes torn and were detained by the crowd. When they were able to escape, they ran through moving traffic[437] to a police station. About that same time the neighbor girl, a child of foreign missionaries, was attacked by Chinese men wielding poles and a crow bar.

Remembering similar anti-foreign incidents in his childhood, Hendon Jr. decided to get his family and his foreign helpers out at once. The Harris family planned to return to Hong Kong after the political situation calmed down and Marjorie could receive adequate medical attention for her health issues. However, their return to Hong Kong never happened.

Because of their need for sudden departure my parents had to buy their tickets for the voyage home on credit. Beverly Emmanuel Baptist Church in Chicago (now Moraine Valley Baptist Church in Palos Heights, Illinois) had sent support to the Harris family for many years.

That church wrote the letter required by the American President Lines guaranteeing that the Harris family's passage back to the United States would be paid. Then that congregation paid that amount out of what was normally sent for the Harrises each month.

On the trip home to America Marjorie led another passenger to faith in Christ.[438]

After they returned to the States the Hendon Harris Jr. family settled in Southern California near their oldest daughter Marjorie Florence. Also a registered nurse, Marjorie had married physician Jack Bettenhausen. They had just started their family – so it was a blessing to my parents to have grandchildren close by. Lillian had married and was teaching at Wheaton College in Illinois. I had just graduated from university and was also living in Illinois.

Hendon Jr. continued returning to Asia frequently. In addition to the work in Hong Kong that was left under the capable care of Chinese workers, the 21 churches in Taiwan continued to be directed by Wen Yeh Ching. Stephen Wu assisted Wen Yeh Ching for many years in that ministry.

Like his grandfather, William Powell, Hendon Harris Jr. spent his later years raising money for mission work. The same year the Harris family left Hong Kong the pastor of their church in that city died so was replaced by another Chinese man.

Hendon Jr. buying antiques to raise money for missions.

In 1970 a medical clinic and Bible school were started in Taitung, Taiwan under the auspices of the Harris's ministry.[439] Later other buildings were added to that campus. The two American student missionaries who had been in Hong Kong eventually married and went to Taiwan and assisted there for a couple of years.

Hendon Jr. took extended trips to Asia and each time spoke in churches or seminaries en route. In 1971 he went into Pakistan to help with relief work among the Bengali refugees there.[440] Later he initiated relief work for refugees from Viet Nam – even going out on boats to rescue those who fled by sea. The finances of his organization were handled by a secretary in the States and underwent the required US audits.

The Divorce

In May 1972 Hendon Jr. and Marjorie divorced.[441] According to the Baptist faith, the divorce disqualified them as missionaries. The divorce was a shock to their children, Hendon and Marjorie's friends, and their supporters.

Not long thereafter Hendon Jr. married a Chinese woman who was more than 30 years younger than himself. That second marriage was unhappy and short lived.

It was shortly after the divorce from my mother that Hendon Jr. found the map for which he is now known – the world map and immediately recognized was associated with the Chinese classic *Shan Hai Jing* and indicated that Chinese mariners reached the Americas at very early dates.

The Map

Hendon Jr. was very excited about the world map and believed that God enabled him to understand it. He wrote in *The Asiatic Fathers of America*:

> We have found a map, which confirms the four-thousand-year–old account of the Shan Hai Jing. In fact, there are at least thirty maps in existence, which vindicate and corroborate the report of the Shan Hai Jing. These world maps, of ancient origin, are found in America, England, France, China, Korea, Japan, and in other countries.
>
> It is now established that the Shan Hai Jing told of at least ten countries in America; which are found on the Harris' Fu Sang [Tian Xia world] map, and on other maps of similar origin, which come from the original Shan Hai Jing map.
>
> So we clearly behold the Shan Hai Jing, the ancient world maps, and the history of the American Indians, together bearing a triple witness that it was the Chinese who discovered and colonized America.
>
> I am grateful to those who loved and aided me, and to the great God of the heavens, without whose guidance all efforts would have been in vain. For in these pages are facts that should, and must bring honor and glory to many people, tongues and nations.
>
> It is my earnest desire that as you consider that lovely long ago morning of Fu Sang, your soul be stirred with a new appreciation of the beauty of life and the innate possibilities of men…that you will hear the music of man's aspirations, deep in your spirit…and catch a glimpse of the excellent countenance of God.
>
> I hope you will have a more sympathetic awareness of men of other persuasions and be drawn to the True Historian who works all things after His will.[442]

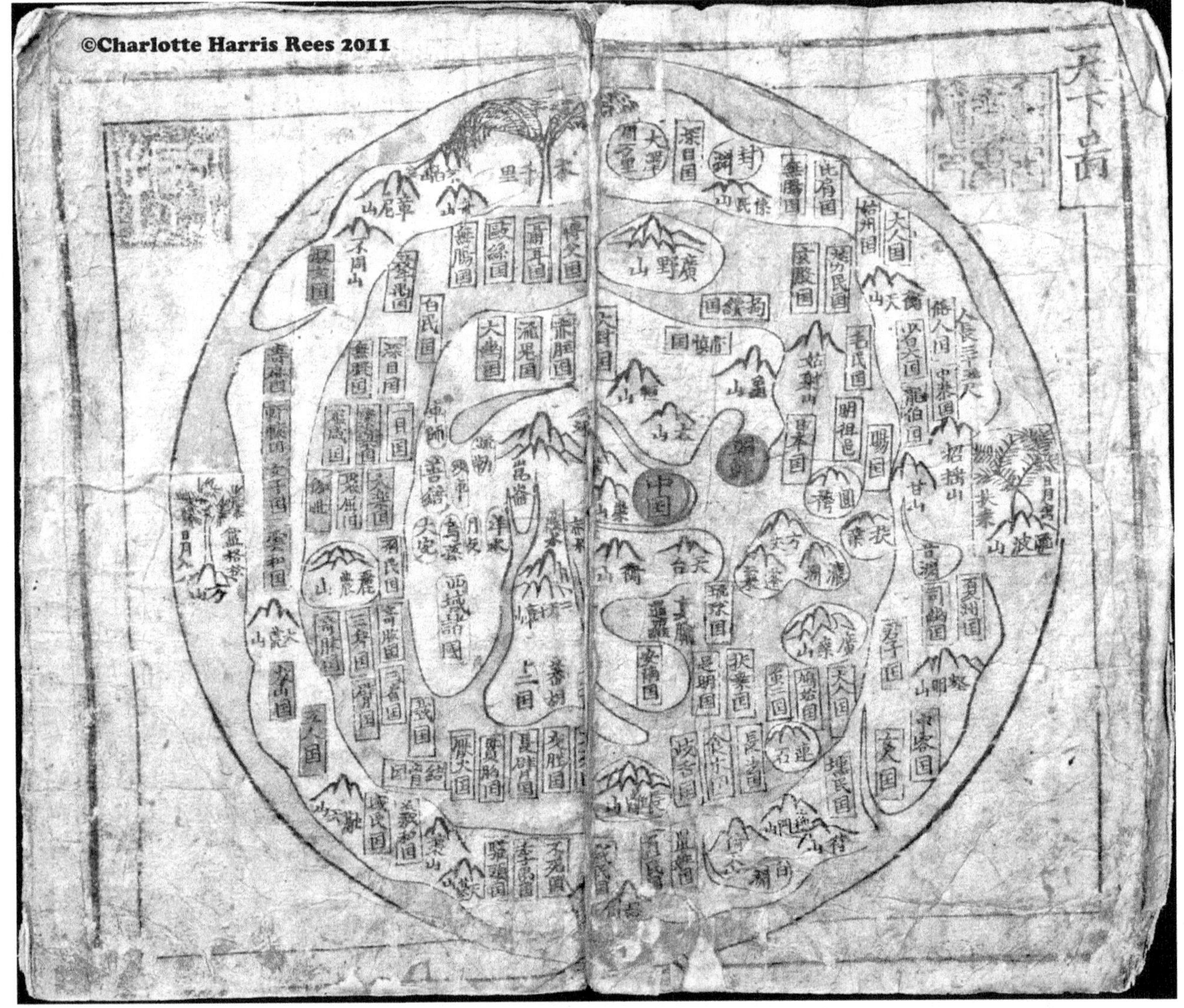

The Tian Xia (Ch'onhado) Map.

The rest of Hendon Jr.'s life was mainly spent researching and talking about the map. He never lost his love of God and continued some humanitarian work. His passports shows that from 1973 through 1980 he continued making trips to numerous countries – Taiwan, Hong Kong, Korea, Belize, South Africa, Thailand, Mexico, Malaysia, Philippines, Singapore, UK, Canada, Australia, and New Zealand.

In January 1981 Father died of a stroke. He was 64 years old. My father was a man before his time – expounding early Chinese exploration of America before many other currently known evidences had appeared. However, Father believed in God and in the maps until the end.

Like Confucius, I believe that my father died believing that his life and attempts to convey his ideas had been a failure.

He had lived his whole life at full throttle. To the end he believed the Scripture that says, "For it is by grace you have been saved through faith and this not from yourselves. It is the gift of God, not by works, so that no one can boast."[443]

He had done many good deeds in life because he was thankful for God's pardon of his own sins and because he loved God and the Chinese people. His good deeds were not to win God's favor. That he already had – despite the fact that he was far from perfect. He received God's forgiveness for his wrongs. He believed: "Christ died for our sins, according to the Scriptures."[444]

My mother died in California in April 2005. A couple weeks later I spoke in a symposium at the Library of Congress about my father's world map.

In 2006 I published an abridgement of Father's *The Asiatic Fathers of America* and since then three other books on various evidences of ancient Chinese exploration in the Americas. Meanwhile, I have spoken about his theories in Europe, Australia, and across Asia and North America – each time mentioning Father.

Though he is now worlds away, does he still "feel my arms around his neck" as he wrote in his poem to me when I was 5 years old?

Books by Dr. Hendon Harris Jr.

Among the books and works written by Dr. Hendon M. Harris Jr:

- *The Asiatic Fathers of America* (two books in one volume), Wen Ho Printing Co., Taipei, 1973
    1. *The Chinese Discovery and Colonization of Ancient America*
    2. *The Asiatic Kingdoms of America*
- *Famous Unwritten Letters*, Light and Hope Publications, Berne, IN, 1956
- *I Predict,* Banchiao, Taiwan, 1952
- *Laughter and Tears,* Light and Hope Publication, Berne, IN, 1954
- *Poems for Grown Up Children*, Light and Hope Publication, Berne, IN
- *The American Idol*, Wabash, IN, 1961
- *The River of Heaven* (a novel set in Taiwan), ca. 1954
- His cantata was titled "The Last Supper."
- His movie was "The Challenge of Formosa."

Addendum 1—

Letter from Foreign Mission Board Rejecting Apostle Paul

This letter below is satire—a playful dig at mission agencies—not criticism of Apostle Paul.

"Foreign Mission Board"

Famous Unwritten Letters
by Hendon M. Harris Jr

Jerusalem, Judea
October 13, 56 A.D.

Rev. Saul Paul
Independent Missionary
Corinth, Greece

Dear Mr. Paul:

We recently received an application from you for service under our Board.

It is our policy to be as frank and open-minded as possible with all our applicants. We have made an exhaustive survey of your case. To be plain, we are surprised that you have been able to "pass" as a bona-fide missionary.

We are told that you are afflicted with a severe eye trouble. This is certain to be an insuperable handicap to an effective ministry. Our Board requires 20-20 vision.

At Antioch, we learn, you opposed Dr. Simon Peter, an esteemed denominational secretary, and actually rebuked him publicly. You stirred up so much trouble at Antioch that a special Board meeting had to be convened in Jerusalem. We cannot condone such actions.

Do you think it seemly for a missionary to do part-time secular work? We hear that you are making tents on the side. In a letter to a church at Philippi you admitted they were the only church to support you. We wonder why.

Is it true that you have a jail record? Certain brethren report that you did two years time at Caesarea and were imprisoned at Rome.

You made so much trouble for the business men at Ephesus that they refer to you as "the man who turned the world upside down." Sensationalism has no place in missions. We also deplore the lurid "over-the-wall-in-a-basket" episode at Damascus.

We are appalled at your obvious lack of conciliatory behavior. Diplomatic men are not stoned and dragged out of the city gate or assaulted by furious mobs. Have you ever suspected that gentler words might gain you more friends? I enclose a copy of Dalius Carnagus' book "How to Win Jews and Influence Greeks."

In one of your letters you refer to yourself as "Paul the aged." It is our settled policy not to accept missionaries over thirty years of age. Furthermore, our new pension policies do not envisage a surplus of super-annuated recipients.

We understand that you are given to fantasies and dreams. At Troas you saw "a man of Macedonia" and at another time "were caught up into the third heaven" and even claimed "the Lord stood by" you. We reckon that more realistic and practical minds are needed in the task of world evangelism.

You have caused much trouble everywhere you have gone. You opposed the honorable women at Berea and the leaders of your own nationality in Palestine. If a man cannot get along with his own people how can he serve foreigners?

We learned that you are a snake-handler. At Malta you picked up a poisonous serpent which is said to have bitten you but you did not suffer harm. Tsk, tsk.

You admit that while you were serving time at Rome that "all forsook you." Good men are not left friendless. Three fine brothers by the names of Diotrephes, Demas, and Alexander, the coppersmith, have notarized affidavits to the effect that it is impossible for them to cooperate with either you or your program.

We know that you had a bitter quarrel with a fellow-missionary named Barnabas. Harsh words do not further God's work.

You have written many letters to churches where you have formerly been pastor. In one of these letters you accused a church member of living with his father's wife, and you caused the whole church to feel badly and the poor fellow was expelled.

You spend too much time talking about "the second coming of Christ." Your letters to the people at Thessalonica were almost entirely devoted to this theme. Put first things first from now on.

Your ministry has been far too flighty to be successful. First Asia Minor, then Macedonia, then Greece, then Italy, and now you are talking about a wild goose chase into Spain. Concentration is more important than dissipation of one's powers. You cannot win the whole world by yourself. You are just one little Paul.

In a recent sermon you said, "God forbid that I should glory in anything save the cross of Christ." It seems to us that you also ought to glory in our heritage, our denominational program, the unified budget and the World Federation of churches.

Your sermons are much too long for the times. At one place you talked until after midnight and a young man was so sleepy that he fell out of the window and broke his neck. Nobody is saved after the first twenty minutes. Stand up, speak up, and then shut up is our advice.

Dr. Luke reports that you are a thin little man, bald, frequently sick, and always so agitated over your churches that you sleep very poorly. He reports that you pad around the house praying half of the night. A healthy mind in a robust body is our ideal for all applicants. A good night's sleep will give you zest and zip so that you wake up full of zing.

We find it best to send only married men into foreign service. We deplore your policy of persistent celibacy. Simon Magus has set up a matrimonial bureau at Samaria where the names of some very fine widows are available.

You wrote recently to Timothy that "You had fought a good fight." Fighting is hardly a recommendation for a missionary. No fight is a good fight. Jesus came not to bring a war but peace. You boast that "I fought with wild beasts at Ephesus." What on earth do you mean?

It hurts me to tell you this, Brother Paul, but in all of my twenty-five years' experience I have never met a man so opposite to the requirements of our Foreign Mission Board. If we accepted you, we would break every rule of modern missionary practice.

Most sincerely yours,

J. Flavius Fluffyhead, Secretary
Foreign Mission Board

JFF/hmh

反共華戰俘

決不回大陸

一美籍教士返臺談

在韓會對戰俘講話

「美聯社臺北十日訊」一個在中國出生的美國浸禮會教士，曾在韓國向所有反共中國籍戰俘講話，要他們不要聽從共黨勸他們回共區的勸告。他現在已經返回臺北。美伊利諾斯州芝加哥人哈里斯博士說：反共中國籍戰俘們似乎決定不回中國大陸去；

但他恐怕有些意志較薄弱的人可能會向壓力屈服。這位曾在芝加哥機場浸禮會教堂担任過牧師的哈氏說：他曾在各營房向中國籍的戰俘作過三十一次的演說，每次的時間爲一小時，總共有一萬三千五百人聽過他的演說。哈氏於三年前，在看見一項他解

釋爲神靈召他協助自由中國的景象之後，即來到臺灣。在他對戰俘的演說中，哈氏保證他們將在臺灣受到熱烈的歡迎。他說：他們可以選擇兩株樹上的果實，一株在臺灣，是自由的果實；另一株在共區，是奴隸的果實。他並向戰俘們唱過一隻他自己所寫的歌──「臺灣之歌」•他說，戰俘們都學會唱這隻歌。他說，每個戰俘身上都刺有「反共抗俄」字樣。他說，他在戰俘營中看見有 蔣總統的背像、青天白日滿地紅的國旗和臺灣地圖。

Wang Tung-yuan

Ambassador Extraordinary and Plenipotentiary

of the Republic of China

Chinese version of article below. In Harris album this article above is photocopied over the card of "Wang Tun-yuan, Ambassador Extraordinary and Plenipotentiary of the Republic of China."

Addendum 2—

"The Prisoners of War Who Opposed the Chinese Communists—Never Return to the Mainland"

News article published in Asia

[Translation of article on previous page from Chinese to English by Naomi Zhang]

An American teacher returned to Taiwan to speak about his conversation which he had with the prisoners of war in Korea. Associated Press, Taipei, on the 10th of this month.

A Chinese-born American Baptist teacher was in Korea to speak with the prisoners of war who opposed the Chinese Communists. He urged them not to listen to the exhortation with which the Communist party is exhorting them to return to the territory of the Communists. He has already returned to Taipei.

Mr. Harris [Hendon Harris Jr.] from Chicago, Illinois said: "It seems like the prisoners of war who opposed the Chinese Communists have decided not to return to the Mainland." But he fears that some of those who are fairly weak in spirit will give in to the pressure.

This Harris, once a pastor in a Baptist church in the Chicago Airport [sic], said that he addressed the Chinese prisoners of war in each of their barracks a total of 31 times, each time speaking for one hour, with a total of 13,500 men hearing him speak.

Three years ago, after Mr. Harris had a vision which he described as God's call on him to assist Free China, he immediately came to Taiwan.

In his speech to the prisoners of war, Mr. Harris guaranteed that they would most certainly receive a warm welcome upon their return to Taiwan. He said that they can choose the fruit from two trees. One is in Taiwan; this is the fruit of freedom. The other is in the Communist's territory; this is the fruit of slavery.

He also sang to the prisoners a song which he wrote – "A Song of Taiwan." He said that the prisoners all learned how to sing this song. He said that all the prisoners had tattooed on their bodies the words: "Oppose the Communists and Soviets." He said that while he was at the Prisoner of War Camp he saw that there was a picture of President Jiang (Chiang Kai Shek), a map of Taiwan, and a flag which possessed a blue sky, a white sun and red land [Nationalist Chinese flag].

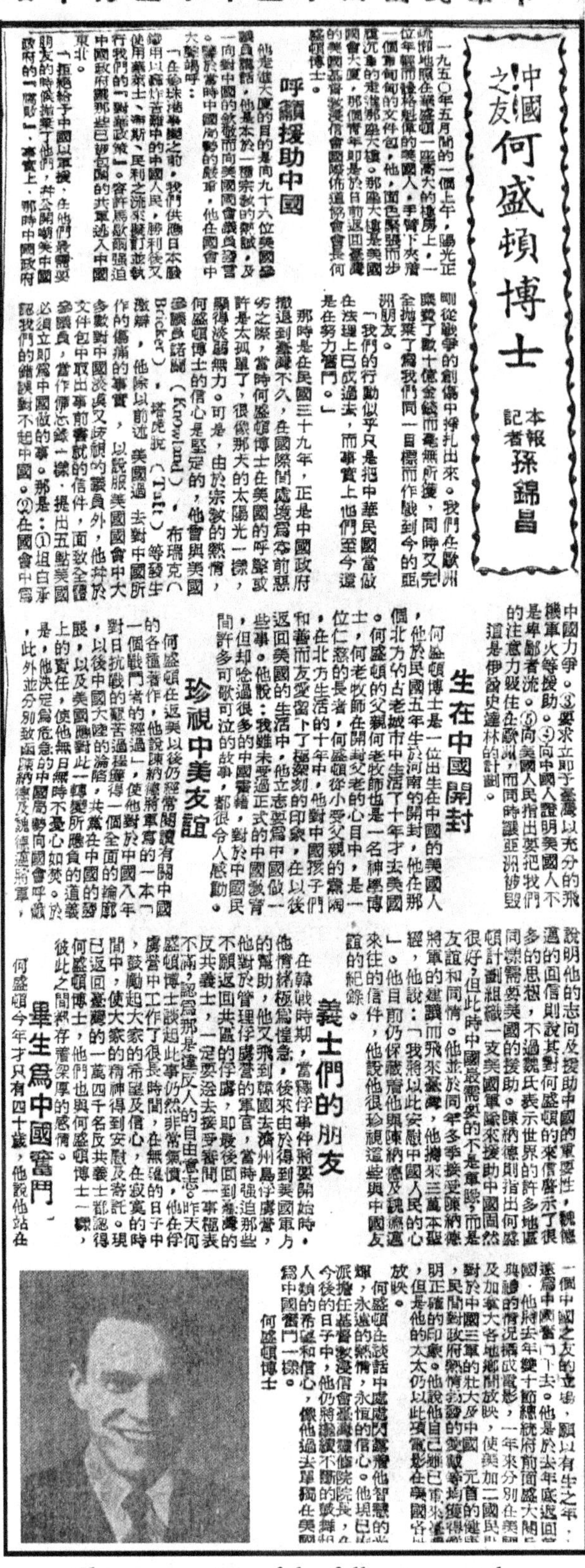

中華民國四十五年十二月十日

中國之友 何盛頓博士

本報記者 孫錦昌

一九五〇年五月間的一個上午，陽光正流瀉地照在華盛頓一座高大的樓房上，一位年輕而體格魁偉的美國人，手臂下夾着一個沉甸甸的文件包，他，面色堅毅而步履沉重的走進那座大樓。那座大樓是美國國會大廈，那個青年即是於日前返回臺灣的美國基督教浸信會國際佈道協會會長何盛頓博士。

呼籲援助中國

他走進大廈的目的是向九十六位美國參議員講話，他是本於一種宗教的熱誠，及一向對中國的欽敬而向美國國會議員發言。鑒於當時中國局勢的嚴重，他在國會中大聲疾呼：

「在珍珠港事變之前，我們供應日本廢鐵，用以轟炸苦難中的中國人民，勝利後又使用蒙來士、海斯、民利之流來擬訂並執行我們的『對華政策』。容許馬歇爾強迫中國政府與那些已被包圍的共軍逃入中國東北。

「拒絕給予中國以軍援，在他們最需要朋友的時候拋棄了他們，并公開嘲笑中國政府的『腐敗』，事實上，那時中國政府剛從戰爭的創傷中掙扎出來。我們在歐洲濫費了數十億金錢而毫無所獲，同時又完全拋棄了爲我們同一目標而作戰到今的亞洲朋友。

「我們的行動似乎只是把中華民國當做在法理上已成過去，而事實上他們至今還是在努力奮鬥。」

那時是在民國三十九年，正是中國政府撤退到臺灣不久，在國際間處境爲空前惡劣之際，當時何盛頓博士在美國的呼聲或許是太孤單了，很像那大的太陽光一樣，顯得淡弱無力。可是，由於宗教的熱情，何盛頓博士的信心是堅定的，他會與美國參議員諾蘭（Knowland）、布瑞克（Bricker）、塔虎脫（Taft）等發生激辯，他除以前述美國過去對中國所作的傷痛的事實，以說服美國國會中大多數對中國淡漠又歧視的議員外，他并於文件包中取出事前書就的信件，面致全體參議員，當作備忘錄一樣，提出五點美國必須立即爲中國做的事。那是：①坦白承認我們的錯誤對不起中國。②在國會中爲中國力爭。③要求立即予臺灣以充分的飛機軍火等援助。④向中國人證明美國人不是卑鄙者流。⑤向美國人民指出要把我們的注意力吸住在歐洲，而同時讓亞洲被毀這是伊喬史達林的計劃。

生在中國開封

何盛頓博士是一位出生在中國的美國人，他於民國五年生於河南的開封，他在那個北方的古老城市中生活了十年才去美國。何盛頓的父親何老牧師也是一名神學博士，何老牧師在開封父老的心目中，是一位仁慈的長者，何盛頓從小受父親的薰陶，在北方生活的十年中，他對中國孩子們和善而友愛留下了極深刻的印象，在以後返回美國的生活中，他立志要爲中國做一些事。他說：我雖未受過正式的中國教育，但卻唸過很多的中國書籍，對於中國民間許多可歌可泣的故事，都很令人感動。

珍視中美友誼

何盛頓在返美以後仍經常閱讀有關中國的各種著作，他說陳納德將軍寫的一本「一個戰鬥者的經過」，使他對於中國八年對日抗戰的艱苦過程獲得一個全面的輪廓，以後中國大陸的淪陷，共黨在中國的發展，以及美國應對此一轉變所應負的道義上的責任，使他無日無時不憂心如焚。於是，他決定爲危急的中國局勢向國會呼籲，此外並分別致函陳納德及魏德邁將軍，說明他的志向及援助中國的重要性，魏德邁的回信則說其對何盛頓的來信啓示了很多的思想，不過魏氏表示世界的許多地區同樣需要美國的援助。陳納德則指出何盛頓計劃組織一支美國軍隊來援助中國固然很好，但此時中國最需要的不是軍隊，而是友誼和同情。他並於同年冬季接受陳納德將軍的建議而飛來臺灣，他攜來三萬本聖經，他說：「我將以此安慰中國人民的心」。他目前仍保藏着他與陳納德及魏德邁來往的信件，他說他很珍視這些與中國友誼的紀錄。

義士們的朋友

在韓戰時期，當釋俘事件將要開始時，他情緒極爲惶急，後來由於得到美國軍方的幫助，他又飛到韓國去濟州島俘虜營，他對於管理俘虜營的軍官，當時強迫那些不願返回共區的俘虜，即最後回到臺灣的反共義士，一定要送去接受審問一事極表不滿，認爲那是違反人的自由意志。昨天何盛頓博士談起此事仍然非常氣憤，他在俘虜營中工作了很長時間，在無望的日子中，鼓勵起大家的希望及信心，在寂寞的時間中，使大家的精神得到安慰及寄託。現已返回臺灣的一萬四千名反共義士都認得何盛頓博士，他們也與何盛頓博士一樣，彼此之間都存着深厚的感情。

畢生爲中國奮鬥

何盛頓今年才只有四十歲，他說他站在一個中國之友的立場，願以有生之年，盡爲中國奮鬥下去。他是於去年底返回美國，他將去年雙十節總統府前盛大閱兵典禮的情況攝成電影，一年來分別在美國及加拿大各地鄉間放映，使美加二國民眾對於中國三軍的壯大及中國元首的健康，民間對政府熱情的愛戴等均獲得鮮明正確的印象。他說他自己雖已重來，但是他的太太仍以此項電影在美國各地放映。

何盛頓在談話中處處閃露着他智慧的光爍，永遠的熱情，永恆的信心。他現已被派擔任基督教浸信會臺灣[illegible]院院長，今後的日子中，他仍將繼續不斷的鼓舞人類的希望和信心，像他過去單獨在美國爲中國奮鬥一樣。

何盛頓博士

Chinese version of the following article.

Addendum 3—

"A Friend of China!—Dr. Hendon Harris [Jr.]"

[Translation of article on previous page from Chinese to English by Naomi Zhang]

Journalist: Sun Jin Chang

The 45th year of the Nationalist Party (1956), December 10

A morning in May, 1950, the sun brightly shining down on a tall building in Washington D.C., a young and strong American with a heavy briefcase tucked under his arm entered this building with a nervous look on his face and heaviness in his step. That building was the Capitol of the United States. And that young man was the president of the American Baptist International Evangelical Society, Dr. Hendon Harris [He Sheng Dun], who just recently returned to Taiwan.

A Call to Help China

His purpose for entering into Congress was in order to speak with 96 American Senators. Being full of religious passion and respect for China, he went to speak with the American Representatives. In view of China's present serious situation, he made this loud call in the midst of the Capitol:

"Before Pearl Harbor, we supplied Japan with scrap iron to use for the purpose of bombing the suffering Chinese people. After the victory, through Wallace (Hua Lai Shi), Hays (Hai Si), and some other notable people, we drafted and executed our 'Policy toward China.'...

"We refused to give military aid to [Nationalist] China, and in the time in which they needed a friend the most we forsook them. Moreover, we publically mocked the "Corruption" of the Chinese government.

"In fact, at that time the Chinese government had just struggled their way out of their wounds from the war. In Europe we wasted several billion dollars and gained nothing. At the same time we completely forsook our Asian friends who have the same purpose with us and who continue to fight until this day.

"Our actions make us appear as though we have considered the Nationalist Party as having already been brought to an end. But actually until now they are still striving heartily."

This was in the 39th year of the Nationalist Party (1950), when the Chinese government had retreated to Taiwan for a short time, they being in the midst of an unprecedentedly harsh international situation.

At that time (in 1950) Dr. Hendon Harris's call in America was perhaps too lonely. But owing to his religious passion Dr. Hendon Harris's faith was firm. He zealously debated with Senators Knowland (Nuo Lan) [now known for helping set US foreign policy priorities and funding regarding Taiwan, later Senate Majority Leader], Bricker (Bu Rui Ke), Taft (Ta Hu Tuo), etc.

He told them all the painful facts about what America had done to China in the past in order to convince the majority of American Congressmen who were indifferent and looked down on China.

He also pulled out of his briefcase a document which he had prepared and presented it to all of the Senators so that it would serve as a memorandum for them. In it he brought out five points concerning the things America must immediately do for [Nationalist] China. Those were as follows:

1. Honestly confess our mistakes to China;
2. Work hard for China in the U.S. Congress;
3. Demand to immediately give enough Airforce munitions, etc. to Taiwan for aid;
4. Prove to the Chinese that Americans are not contemptable people;
5. Point out to the American people that to focus our attention on Europe while at the same time allowing Asia to be destroyed is the very plan of Stalin the leader of Russia.

Born in Kaifeng, China

Dr. Hendon Harris is an American who was born in China. He was born in Kaifeng, Henan during the fifth year of the Nationalist Party rule (1916). It was not until he had lived in that ancient northern city for ten years that he finally went to America.

Hendon Harris's father, Pastor Harris, was also a Doctor of Theology. In the eyes of the people of Kaifeng, Pastor Harris was a compassionate elder. Hendon Harris [Jr.] from a young age was influenced by his father. While living in northern China for those ten years, he received a deep impression from the kind and friendly children.

In the time of his life after returning to America, he determined to do some things for China. He said: "Although I never received formal Chinese education, I did read many Chinese books concerning both happy and tragic stories of the Chinese people, all of which were very moving."

Treasuring this Chinese-American Friendship

Hendon Harris, after returning to America, still often read all kinds of works related to China. He said the book entitled *Way of a Fighter* which General Claire Chennault (Chen Na De) wrote caused him to understand a rough sketch of the bitter process of China's eight year war with Japan, the later fall of mainland China into enemy hands, the development of the Communist Party in China, and the morally appropriate response America should have to this change, all of which caused him to be unceasingly filled with grief and sorrow.

Therefore, he determined to call out to Congress in support of China who is in a situation of crisis. Moreover, he also wrote a letter to Generals Chennault and Wedemeyer (Wei De Mai), explaining his ambition and the importance of assisting China.

The letter of response from Wedemeyer said that the letter he received from Hendon Harris caused him much thought. General Wedemeyer, however, expressed that there are many regions in the world which likewise need American aid.

Chennault pointed out that Hendon Harris's plan to organize a U.S. Army to assist China is indeed a good plan, but at this time what China needs the most is not an army but friendship and sympathy.

In the winter of that year, Harris accepted the suggestion of General Chennault and returned to Taiwan. Bringing with him 30,000 Bibles, he said: "I will comfort the hearts of the Chinese people with these."

Unto this day, he still holds on to the letters written between Chennault, Wedemeyer, and himself. He said he highly values these records concerning our friendship with China.

A Friend of Righteous People

During the war, right as the incident of the release of captives was about to start, he was emotionally panicked. Later on, owing to the help of the U.S. Army which he received, he flew again to the prison camp which is located in Jeju Island, Korea.

He expressed his dissatisfaction concerning the military officers who were in charge of managing the prison camp, who were persecuting those prisoners who refused to return to the Communist territory and also those who finally returned to Taiwan's "Righteous Opponents of the Communist Party," as they were being forced by these officers to be interrogated. He considers that to be a violation of man's free will.

Yesterday as Dr. Hendon Harris spoke of this matter, he was still very indignant. He had worked in the prison camp a long time, and during those hopeless days he encouraged everyone's hope and faith. During this lonely time, he caused everyone's spirit to be comforted and to receive sustenance.

Now all 14,000 of those "Righteous Opponents of the Communist Party" who have returned to Taiwan know Dr. Hendon Harris. They are also like Dr. Hendon Harris, in that they have a deep relationship with one another.

A Lifelong Fight for China

This year, Hendon Harris is but 40 years old. He said that he stands as a friend of China, and desires to fight for China for the rest of his life. He returned to America at the end of last year. He filmed and made a movie of last year's magnificent parade of soldiers which took place in front of the President's Palace on Taiwan National Day (Shuang Shi Jie), October 10th.

Within one year, in both America and Canada, he played the film in their towns, in order to cause America and Canada, these two nations, to get a fresh, clear, and correct impression of the greatness of the three branches of the [Nationalist] Chinese military, the health of the [Nationalist] Chinese heads of state, and the enthusiastic love of the [Nationalist] Chinese people for their government. He said that he himself had already come back to Taiwan, but his wife is still in the towns of America showing this movie.

While Hendon Harris was speaking, the light of his wisdom, his eternal passion, and his everlasting faith, shown everywhere. Currently, he has already been commissioned to take position as the president of the Taiwan Christian Baptist Seminary. In the days ahead, he will still be unceasingly encouraging the hope and faith of humanity, just as in the past when in American he fought alone for China.

Acknowledgements

Thanks to all who helped in the production of this book. You are appreciated. Special thanks go to: Robert Alderman; Jim Berwick (Southern Baptist Archives; Richmond, VA); Barbara Blann; George Blann; James Gillespie; Hendon Harris III (photos and correspondence); John Harris (photos); Larry Harris (photos); Lawrence Harris (photos); Bethany Hawkins; Aurora Harris Heller (photos); Robert Heller; Hwa-Wei Lee; Library Staff, Bedford County Virginia; Michael Lowry; Jenny Manasco (Union University); Ron Merkle; Buford Nichols Jr. (photos); John Nichols; Dave Rees; Juanita Rees (missionary prayer letters); Elaine Rice; Harty Schmaehl; Marjorie Florence Harris Schmaehl; Florencita (Cita) Harris Strunk; Peter Stursberg; Wally Turnbull; Roger Voegtlin; Ralph Wong; Naomi Zhang.

Six of the seven Harris siblings in 2013: L to R Aurora, John, Hendon III, Charlotte, Lillian, Marjorie.

The seven Harris siblings in 2015: L to R Marjorie, Lillian, Charlotte, Hendon III, John, Aurora, and Mejchahl.

About Charlotte Harris Rees

I feel very blessed for opportunities that have come my way including to appear on television and National Public Radio in the United States and Canada and in numerous international news articles. From 2005 on I have given historical presentations including at the Library of Congress (Washington, DC); National Library of China (Beijing); Royal Geographical Societies (London and Hong Kong). La Trobe University and University of Adelaide, (Australia); Seton Hall; Stanford University; Tsinghua and Peking Universities (Beijing); University of London; University of British Columbia; University of Hawaii; Portland State University (Oregon); Macau University; the University of Maryland; Liberty University; at the Chinese Historical Society (Los Angeles); in Switzerland; and at symposiums in Melaka, Malaysia; Washington, DC; Victoria, BC; and Shanghai.

My previous books include *Did Ancient Chinese Explore America? My Journey Through the Rocky Mountains to Find Answers* (2013), *New World Secrets on Ancient Asian Maps* (2011, 2014), and *Secret Maps of the Ancient World* (2008, 2009). In 2006 my abridged version of my father's, *The Asiatic Fathers of America: Chinese Discovery and Colonization of Ancient America* was published. The World Confederation of Institutes and Libraries for Chinese Overseas Studies (WCILCOS) lists my books.

My father, Dr. Hendon M. Harris Jr. (1916-1981), a third generation Baptist missionary, was born in Kaifeng, China. In 1972 he found in an antique shop in Korea an ancient Asian map which led him to write a book of almost 800 pages that contended early arrival of Chinese to America by sea. In 2003 my brother, Hendon III, our spouses, and I took the Harris Map Collection to the Library of Congress where it remained for three years while being studied.

A graduate of Columbia International University, I am an independent researcher. Dr. Cyclone Covey, History Professor Emeritus, Wake Forest University, (PhD from Stanford), who for over 60 years studied the Chinese connection to very early American history, was my research mentor from 2003 until his death in late 2013. For more than 10 years Dr. Hwa-Wei Lee, retired Chief of the Asian Division of the Library of Congress, has befriended me and endorsed my books.

As a child I lived for four years in Taiwan then later in Hong Kong where my parents served. In recent years I have made several trips to China. Parents of three, but now empty nesters, my husband, Dave Rees and I live in Virginia. My web site is www.AsiaticFathers.com and e-mail address HarrisMaps@msn.com.

Notes

1. "The Families of Samuel Rankin and James Rankin," Luginbuel Funeral Home, Chapter 4, Fifth Generation, The family of Nancy Ann Rankin and Wm. Madison Powell web 4 Dec. 2006 www.assets.luginbuel.com/genealogy/documents/Rankin,%20James%20Samuel%Family.pdf.
2. I. G. Murray, "W.D. Powell, the Champion Church Dedicator," *Baptist and Reflector,* Vol. 96, No. 43, Nashville, TN, October 23, 1930, p. 1.
3. "The Families of Samuel Rankin and James Rankin," Luginbuel Funeral Home, Chapter 4, Fifth Generation, The family of Nancy Ann Rankin and Wm. Madison Powell web 4 Dec. 2006 www.assets.luginbuel.com/genealogy/documents/Rankin,%20James%20Samuel%Family.pdf.
4. Florence Powell Harris, *How Beautiful the Feet,* Hong Kong: Luen Shing Printing Co., 1968, p. 8-9.
5. Murray, p. 1.
6. "The Families of Samuel Rankin and James Rankin," Luginbuel Funeral Home, Chapter 4, Sixth Generation, The family of Rev. William David Powell and Mary Florence Maberry web www.assets.luginbuel.com/genealogy/documents/Rankin,%20James%20Samuel%Family.pdf.
7. Florence Powell Harris, p. 9.
8. Ibid.
9. Florence Powell Harris, pp. 9 – 10.
10. James Alex Baggett, *So Great a Cloud of Witnesses: Union University 1823-2000,* Jackson, TN: Union University Press, 2000, pp. 33, 50.
11. Murray, p. 1.
12. W. D. Powell as told to Rupert Richardson, "A Baptist Preacher on the Texas Frontier," *West Texas Historical Association Year Book,* Vol. IX, October 1933, p. 48.
13. Baggett, pp. 48, 54.
14. Baggett, pp. 66-67.
15. James Garvin Chastain, *Thirty Years in Mexico,* El Paso, TX: Baptist Publishing House, 1927, p.175.
16. Murray, p. 1.
17. W. D. Powell as told to Rupert Richardson, p. 48.
18. George R. Jewell, "W. D. Powell, Traveling Secretary of Foreign Board, Dies in Alabama," *Western Recorder,* Vol. 108, No. 21, Louisville, Kentucky, May 24, 1934, p. 20.

19. L. S. Foster, *Mississippi Baptist Preachers,* St. Louis, MO: National Baptist Publishing Co., 1895, p.553.
20. W. D. Powell, "Letter to Luther M. Vaughter," October 1, 1930, p. 1 (now at archives@ rutherfordcountytn.gov).
21. Foster, p. 553.
22. Bethany L. Hawkins, "Powell's Chapel: A Church and Community," Master's Thesis, Middle Tennessee State University, August 2014, p. 7-8, quoting Oct. 1, 1930 letter from Wm. Powell to Vaughter.
23. Hawkins, p. 9, quoting Oct. 1, 1930 letter from Wm. Powell to Vaughter.
24. W. D. Powell, Letter to Vaughter, p. 1.
25. W. D. Powell, Letter to Vaughter, p. 2.
26. "Powell's Chapel Baptist Church," n.d. web 25 Mar. 2016 http://www.powellschapel.com/ Church's_History/.
27. "Descendants Database Search," Ancestor A094379, nd Web 4 Dec 2011 http://www.services. dar.org/public/dar_research/search_desendants/?action=list&My….
28. Florence Powell Harris, p.15.
29. "The Families of Samuel Rankin and James Rankin," Luginbuel Funeral Home, Chapter 4, Fifth Generation, The family of Nancy Ann Rankin and Wm. Madison Powell web 4 Dec. 2006 www.assets.luginbuel.com/genealogy/documents/Rankin,%20James%20Samuel%Family. pdf.
30. "The Families of Samuel Rankin and James Rankin," Luginbuel Funeral Home, Chapter 4, Fifth Generation, The family of Nancy Ann Rankin and Wm. Madison Powell web 4 Dec. 2006 www.assets.luginbuel.com/genealogy/documents/Rankin,%20James%20Samuel%Family. pdf.
31. Florence Powell Harris, p. 10.
32. "William David Powell," # 234, Operation Baptist Biography Data Form for Deceased Person, F. M. B., 9/25/59, #15
33. "The Families of Samuel Rankin and James Rankin," Luginbuel Funeral Home, Chapter 4, Fifth Generation, The family of Nancy Ann Rankin and Wm. Madison Powell web 4 Dec. 2006 www.assets.luginbuel.com/genealogy/documents/Rankin,%20James%20Samuel%Family. pdf.
34. W. D. Powell, Letter to Vaughter, p. 1.
35. Alexi Assmus, "Early History of X-rays" n.d. Web 4 April 2016 http://www.slac.stanford.edu/ pubs/beamline/25/2/25-2-assmus.pdf.
36. Ananya Mandal, MD, "History of Tuberculosis," *News Medical Life Sciences & Medicine* n.d. Web 27 April 2016 http://www.news-medical.net/health/History-of-Tuberculosis.aspx.
37. Edine W. Tiemersma, Marieka J. van der Werf, Martien W. Borgdorff, Brian G. Williams, and Nico J. D. Nagelkerke, "Natural History of Tuberulosis: Duration and Fatality of Untreated Pulmonary Tuberculosis," *PLOS,* http://www/ncbi.nlm.nih.gov/pmc/articles/PM3070694/.
38. Ibid.
39. W. D. Powell as told to Rupert Richardson, pp. 48-59.
40. W. D. Powell as told to Rupert Richardson, pp. 48-51.

41. W. D. Powell as told to Rupert Richardson, p. 51.
42. Justice C. Anderson, *An Evangelical Saga,* Maitland, FL: Xulon Press, 2005, p. 91.
43. W. D. Powell as told to Rupert Richardson, p. 52.
44. Jewell, p. 20.
45. W. D. Powell as told to Rupert Richardson, pp. 48, 52.
46. Murray, p. 1.
47. Timothy Larsen, "When Did Sunday Schools Start," Christian History, n.d. web 30 Mar 2016 http://www.christianitytoday.com/history/2008/august/when-did-sunday-schools-start.html. pp. 1-2.
48. Ibid.
49. W. D. Powell as told to Rupert Richardson, p. 54.
50. W. D. Powell as told to Rupert Richardson, p. 56.
51. W. D. Powell as told to Rupert Richardson, p. 57.
52. Charlotte H. Rees, *Did Ancient Chinese Explore America? My Journey Through the Rocky Mountains to Find Answers,* Durham, NC: Torchflame Books, 2013, pp. 114-115.
53. W. D. Powell as told to Rupert Richardson, p. 58.
54. W. D. Powell as quoted by Benjamin Franklin Fuller, *History of Texas Baptists,* Louisville, KY: Baptist Book Concern, 1900, pp. 258-259.
55. Fuller, p. 259.
56. "Mexico Missions," nd web 22 March 2016 www.archives.imb.org/images/Mexico,%201889-1902.pdf, p. 34.
57. "Mexico Missions," nd web 22 March 2016 www.archives.imb.org/images/Mexico,%201889-1902.pdf, p. 3.
58. Justice C. Anderson, "Nineteenth Century Baptist Missions in Mexico, 1827 -1903," *Baptists and Mission: Papers from the Fourth International Conference on Baptist Studies,* Studies in Baptist History and Thought, Vol. 29, Eugene, Oregon: Wipf & Stock, 2008, p. 159.
59. Anderson, *Baptists and Mission,* p. 151.
60. Anderson, *An Evangelical Saga,* p. 91.
61. W. D. Powell, "Laying Foundations in Modern Mexico," J. N. Prestridge, *Modern Baptist Heroes and Martyrs,* Louisville, KY: The World Press, 1911, pp. 270-271.
62. Frank W. Patterson, *A Century of Baptist Work in Mexico,* El Paso, TX: Baptist Spanish Publishing House, 1979, p. 44.
63. Anderson, *Baptists and Mission:* p. 160.
64. A. T. Hawthorne, Letter to H. A. Tupper, SBFMB, March 3, 1882.
65. W. D. Powell, letter to Robert J. Willingham, Secretary of Foreign Mission Board, Southern Baptist Convention, Feb. 17, 1897 in archives of Southern Baptist Convention.
66. W. D. Powell as told to Rupert Richardson, p. 58.
67. Murray, p. 1.
68. Murray, p. 4.
69. Chastain, p. 128.
70. Murray quoting Powell, p. 4.

71. "Mexico Missions," nd web 22 March 2016 www.archives.imb.org/images/Mexico,%201889-1902.pdf, p. 35.
72. "Mexico Missions," nd web 22 March 2016 www.archives.imb.org/images/Mexico,%201889-1902.pdf, p. 11.
73. Patterson, p. 46.
74. Powell, "Laying Foundations," Prestridge, pp. 270-271.
75. "Mexico Missions," nd web 22 March 2016 www.archives.imb.org/images/Mexico,%201889-1902.pdf, p. 9.
76. "Mexico Missions," nd web 22 March 2016 www.archives.imb.org/images/Mexico,%201889-1902.pdf, p. 10.
77. "Mexico Missions," nd web 22 March 2016 www.archives.imb.org/images/Mexico,%201889-1902.pdf, p. 11.
78. Florence Powell Harris, p. 11.
79. Powell, "Laying Foundations," Prestridge, p. 274.
80. Murray, p. 1 quoting Powell.
81. Powell, "Laying Foundations," Prestridge, p. 272.
82. Patterson, p. 46.
83. "Mexico Missions," nd web 22 March 2016 www.archives.imb.org/images/Mexico,%201889-1902.pdf, p. 9.
84. Chastain, p. 129.
85. Chastain, p. 128.
86. Patterson, pp. 49-50.
87. Chastain, pp. 105, 129.
88. Foster, p. 553.
89. Chastain, p. 99.
90. Chastain, p. 104.
91. Patterson, p. 67.
92. Anderson, *An Evangelical Saga,* p. 94.
93. Anderson, *An Evangelical Saga,* p. 96.
94. Florence Powell Harris, p. 12.
95. Patterson, p. 51.
96. New King James version of the Bible.
97. Powell, "Laying Foundations," Prestridge, pp. 274-277.
98. "Mexico Missions," nd web 22 March 2016 www.archives.imb.org/images/Mexico,%201889-1902.pdf, p. 36.
99. Powell, "Laying Foundations," Prestridge, p. 272.
100. "Mexico Missions," nd web 22 March 2016 www.archives.imb.org/images/Mexico,%201889-1902.pdf, p. 19.
101. Murray, p. 1.

102. "Mexico Missions," nd web 22 March 2016 www.archives.imb.org/images/Mexico,%201889-1902.pdf, p. 28.
103. Ibid.
104. Patterson, p. 47 citing *Records, ABHMS,* Vol.10, p. 191.
105. "Mexico Missions," nd web 22 March 2016 www.archives.imb.org/images/Mexico,%201889-1902.pdf, p. 9.
106. Chastain, p. 130.
107. "Mexico Missions," nd web 22 March 2016 www.archives.imb.org/images/Mexico,%201889-1902.pdf, p. 36.
108. Powell, "Laying Foundations," Prestridge, pp. 272 -273.
109. "William David Powell," # 234, Operation Baptist Biography Data form for Deceased Person, F. M. B., 9/25/59.
110. Chastain, p. 165.
111. Patterson, pp. 53,54,59,61.
112. Patterson, p. 54.
113. Florence Powell Harris, p. 4.
114. Florence Powell Harris, p. 4-5.
115. Chastain, p. 172.
116. "Miss Anna J. Maberry," *The Foreign Mission Journal,* Vol. XXIV – December, 1892-No. 5, pp. 131-132.
117. Chastain, p. 165.
118. W. D. Powell, letter to Dr. Willingham dated Dec. 13, 1895 now in archives of Southern Baptist Convention.
119. Ibid.
120. Chastain, p. 134.
121. Benjamin Riley, *History of the Baptists of Texas,* Dallas, 1907, p. 315.
122. W. D. Powell, letter to Willingham, August 28, 1898, in archives of Southern Baptist Convention.
123. "Mexico Missions," nd web 22 March 2016 www.archives.imb.org/images/Mexico,%201889-1902.pdf, p. 51.
124. Ibid.
125. Ibid.
126. "Mexico Missions," nd web 22 March 2016 www.archives.imb.org/images/Mexico,%201889-1902.pdf, p. 63.
127. W. D. Powell, letter to Willingham, Feb. 15, 1897, in archives of Southern Baptists.
128. Chastain, p. 132.
129. Foster, p.556.
130. Anderson, *Baptists and Mission,* p. 161.
131. Florence Powell Harris, p. 18.

132. "William Jennings Bryan," nd web 21 Jul 2016 www.http://projects.vassar.edu/1896/bryan.html.
133. Edward H. Worthen, "The Mexican Journeys of William Jennings Bryan, A Good Neighbor," pp. 485-486. www.NebraskaHistory.org,
134. Chastain, p. 134.
135. W. D. Powell, letter to Willingham, April 10, 1894, in archives of Southern Baptist Convention.
136. Chastain, p. 132-134.
137. W. D. Powell, letter to Willingham, April 8, 1895, now in archives of Southern Baptists.
138. W. D. Powell, letter to Willingham, February 14, 1894, in archives of Southern Baptists.
139. Photo courtesy of W. D. Powell facebook page.
140. Chastain, p. 132.
141. W. D. Powell, letter to H. H. Harris, August 27, 1894, in archives of Southern Baptists.
142. W. D. Powell, letter to Willingham, April 30, 1895, in archives of Southern Baptists.
143. W. D. Powell, letter to Willingham, May 3, 1895, in archives of Southern Baptists.
144. W. D. Powell, letter to Willingham, June 15, 1895, in archives of Southern Baptists.
145. W. D. Powell, letter to Willingham, June 19, 1895, in archives of Southern Baptists.
146. W. D. Powell, letters to Willingham, June 15, 1895, June 16, 1895, letter from D. L. Moody, dated June 4, 1895 in archives of Southern Baptists.
147. W. D. Powell, letter to Willingham, July 2, 1895, in archives of Southern Baptists.
148. W. D. Powell, letter to Willingham, Nov. 18, 1895, in archives of Southern Baptists.
149. W. D. Powell, letter to Willingham, December 2, 1895.
150. Ibid.
151. W. D. Powell, letter to Willingham, April 9, 1896.
152. W. D. Powell, letter to Willingham, May 1, 1896.
153. W. D. Powell, letter to Willingham, May 25, 1896.
154. W. D. Powell, letter to Willingham, July 13, 1896.
155. Ibid.
156. W. D. Powell, letter to Willingham, October 30, 1896.
157. W. D. Powell, letter to Willingham, December 31, 1896.
158. Thomas Crittendon, letter to William Powell, January 30, 1897 in Southern Baptist archives.
159. "Missions Row," *Mexican Herald*, n. d.
160. Various letters from 1897 in archives of Southern Baptists.
161. Alejandro Trevino to Willingham, March 3, 1897.
162. W. D. Powell, letter to Willingham, second letter of Feb. 15, 1897.
163. W. D. Powell, letter to Willingham, Feb. 17, 1897.
164. W. D. Powell, letter to Willingham, February 26, 1897.
165. Florence Powell Harris, *How Beautiful the Feet,* reprint by Larry Harris, March 2011, p. 289.
166. W. D. Powell, letter to Willingham, March 25, 1897.
167. W. D. Powell, Letter to Willingham, Feb. 17, 1897.

168. W. D. Powell, letter to Willingham, April 10, 1897.
169. W. D. Powell, letter to Willingham, Oct. 4, 1897.
170. W. D. Powell, letter to Willingham, Nov. 22, 1897.
171. W. D. Powell, letter to Willingham, January 3, 1898.
172. W. D. Powell, letter to Willingham, January, 1898.
173. Ibid.
174. W. D. Powell, letter to Willingham, Feb. 28. 1898.
175. "Mexico Missions," nd web 22 March 2016 www.archives.imb.org/images/Mexico,%201889-1902.pdf, p. 73.
176. W. D. Powell, letter to Willingham March 20, 1898.
177. W. D. Powell, letter to Willingham, March 28, 1898.
178. W. D. Powell, letter to Willingham, April 4, 1898.
179. W. D. Powell, letter to Willingham, May 6, 1898.
180. W. D. Powell, letter to Willingham, May 25, 1898.
181. W. D. Powell, letter to Willingham, May 27, 1898.
182. W. D. Powell, letter to Willingham, June 2, 1898.
183. W. D. Powell, letter to Willingham, Dec. 17, 1909.
184. Powell, as quoted by Murray, pp. 1 and 4.
185. Anderson, *An Evangelical Saga,* p. 93.
186. Benjamin Franklin Riley, *A History of Baptists in Southern States East of the Mississippi,* Philadelphia: American Baptist Publication Society, 1898, pp. 236-238.
187. Anderson, *Baptists and Mission,* p. 160.
188. Murray, quoting Powell, p. 4.
189. Murray, quoting Powell, p. 4.
190. Baggett, p. 75.
191. Baggett, photos before p. 71.
192. Baggett, p. 60.
193. Baggett, p. 90.
194. Baggett, Illustration 3.4.
195. Murray, p. 4
196. Ibid.
197. Ibid.
198. Photo courtesy of W. D. Powell Facebook page.
199. Ibid.
200. Jewell, p. 20.
201. "The Families of Samuel Rankin and James Rankin," Luginbuel Funeral Home, Chapter 4, Sixth Generation, The family of Rev. William David Powell and Mary Florence Maberry web www.assets.luginbuel.com/genealogy/documents/Rankin,%20James%20Samuel%Family.pdf.
202. "Marquis Who's Who On Demand," n.d. web 19 March 2016 www.https;//cgi.marquiswhoswho.com/OnDemand/Default.aspx?last_name=Powell&first_name=William.

203. Phone call with Cita Strunk on 29 March 2016.
204. Patterson, p. 7.
205. Ibid.
206. "William David Powell," Operation Baptist Biography Data Form for Deceased Person, question 2.
207. "Descendants Database Search," Ancestor A094379, nd Web 4 Dec 2011 services.dar.org/public/dar_research/search_desendants/?action=list&My....
208. "Descendants Database Search," Ancestor A094379, nd Web 4 Dec 2011 services.dar.org/public/dar_research/search_desendants/?action=list&My....
209. Jewell, p. 20.
210. Murray, p. 4.
211. Chastain, p. 175.
212. Florence Powell Harris, p. 16.
213. E-mail dated 9/16/2016 from Larry Harris who has worked extensively on our family genealogy.
214. "Robert Morrison: 1782 -1834," n.d. web 20 Feb 2016, http://www/bdccpm;ome/meten/stories/m/morrison-robert.php.
215. Ibid.
216. Paul Hattaway, *Henan: Fire & Blood*, Vol. 2, Carlisle, UK: Piquant Editions, 2009, p. 49.
217. Hattaway, p. 51.
218. Ibid.
219. Hattaway, p. 48.
220. *Annual*, Southern Baptist Convention, 1913, p. 184.
221. Hattaway, p. 86.
222. Hattaway, p. 85.
223. Hattaway, p. 86.
224. Hattaway, p. 70-73.
225. "Making disciples and multiplying churches," n. d. Web 22 April 2016 http.//www.imb.org/about-us/history.aspx#.VxoFWDArldU.
226. *Annual*, Southern Baptist Convention, 1915, p. 194.
227. John Keay, *China: A History*, New York: Basic Books, Harper Collins, 2009, p. 499.
228. Keay, p. 500.
229. Keay, p. 493.
230. Hattaway, p. 43.
231. Annie Jenkins Sallee, *W. Eugene Sallee: China's Ambassador*, Nashville, TN: The Sunday School Board of the Southern Baptist Convention, 1933, p. 10.
232. Florence Powell Harris, p. 47.
233. According to Funk and Wagnalls Encyclopedia, "Late in the 12th Century a colony of Jews settled in the city. The colonists maintained their racial purity and religious faith for many centuries, but were gradually assimilated through intermarriage." See picture of my grandfather with some of these Jewish Chinese. They even left a written history giving the names of the immigrants.

234. James R. Gillespie, *Around the World and Headed South: Growing Up as a Twin and a Missionary Kid*, Third Edition, 2003, p. 4.
235. *Journey Into China,* National Geographic staff, Seattle: U of Washington Press, 1982, p. 132.
236. "Marco Polo," nd web 31 Mar 2016 http://www.chinaculture.org/gb/en_aboutchina/2003-09/24/content_22629.htm.
237. Sallee, p. 58.
238. Sallee, p. 75.
239. Hendon Harris, "Indigenous Churches in China," doctoral thesis for Southern Baptist Seminary, Louisville, Ky, 1927, p. 165 – 166.
240. Keay, p. 481.
241. Charles Maddry, *Christ's Expendables,* Nashville, TN: Broadman Press, 1949, p. 17.
242. Hendon Harris, "Indigenous Churches in China," p. 70-71.
243. Hendon Harris, "Indigenous Churches in China," p. 47.
244. *Annual,* Southern Baptist Convention, 1912, p. 236.
245. Florence Powell Harris, p. 150.
246. Florence Powell Harris, p. 51.
247. *Annual,* Southern Baptist Convention, 1912, p. 232.
248. *Annual,* Southern Baptist Convention, 1912, p. 231.
249. *Annual,* Southern Baptist Convention, 1912, p. 95.
250. *True Spiritual Roots for All Chinese,* Arcadia, CA: Shangti-research.org, p. 28.
251. New Kings James version of the Bible.
252. Florence Powell Harris, p. 94.
253. Florence Powell Harris, p. 65-67.
254. Florence Powell Harris, p. 68.
255. Sallee, p. 93.
256. Florence Powell Harris, p. 93.
257. Sallee, p. 196.
258. Florence Powell Harris, p. 97.
259. *Annual,* Southern Baptist Convention, 1914, pp. 226-227.
260. Erleen J. Christensen, *In War and Famine: Missionaries in China's Honan Province in the 1940's,* Montreal: McGill-Queen's University Press, 2005, p. 18.
261. *Annual,* Southern Baptist Convention, 1915, p. 194.
262. Florence Powell Harris, p. 71.
263. Hendon Harris, "Indigenous Churches in China," p. 165- 166.
264. Florence Powell Harris, p. 74.
265. *Ann*ual, Southern Baptist Convention, 1912, p. 231.
266. "The Chinese New Year of Dragon," n.d. web 7 January 2010 http://www.index-china.com/index-english/Dragon-year.html.
267. Florence Powell Harris, p. 74.
268. Florence Powell Harris, p. 75.

269. Rebekah E. Adams, *Called to China: Attie Bostick's Life and Missionary Letters from China: 1900 – 1943,* Huntsville, AL: Halldale Publishing Co., 2006, p. 20.
270. "Addie Cox," Stories: *Biographical Dictionary of Chinese Christianity,* n.d. web Feb. 20, 2016, http://www.bdcconline.net/enstories/c/coc-addie.php.
271. Ibid.
272. Hendon Harris, "Indigenous Churches in China," p. 179.
273. Charles W. Hayford, *To the People: James Yen and Village China,* New York: Columbia U Press, 1990, p. 24.
274. "Hendon Mason Harris," #2606, Operation Baptist Biography Data Form for Living Person.
275. Florence Powell Harris, p. 84.
276. Ibid.
277. "The Influenza Pandemic of 1918," n.d. Web 12 February 2016 http://wwwvirus.stanford.edu/uda/.
278. Florence Powell Harris, p. 83.
279. Florence Powell Harris, p. 82.
280. Florence Powell Harris, p. 88.
281. Hattaway, p. 69-70.
282. Sallee, p. 201.
283. Peter Stursberg, *No Foreign Bones in China,* Alberta, BC: The University of Alberta Press, 2002, p. 142.
284. Hattaway, p. 62-64.
285. Stursberg, p. 147-149.
286. Hattaway, p. 64.
287. Stursberg, p. 152.
288. Florence Powell Harris, p. 91.
289. Florence Powell Harris, p. 127.
290. Florence Powell Harris, p. 128.
291. Maddry, p. 164.
292. Sallee, pp. 163 -168.
293. Sallee, p. 143-144.
294. Florence Powell Harris, p. 142.
295. Floreence Powell Harris, p. 142
296. Annie Jenkins Sallee, *Torchbearers in Honan*, Nashville: Broadman Press, 1948, pp. 67-68.
297. Sallee, pp. 240-241.
298. Florence Powell Harris, p. 149.
299. Florence Powell Harris, p. 152.
300. Florence Powell Harris, p. 158.
301. Per phone conversation with Cita on 9/ 21, 2016.
302. Florence Powell Harris, p. 176.
303. Christensen, p. 38.

304. *Through Fire, The Story of 1938,* London: China Inland Mission, 1939, p. 19.
305. This is not included in my grandmother's book, but has long been told in my family. On 14 March 2016 I confirmed it over the phone with my Aunt Cita Strunk who said that it was true. I have no other way to confirm.
306. Florence Powell Harris, p. 176.
307. Hattaway, p. 89.
308. *Through Fire,* p. 19.
309. *Annual,* Southern Baptist Convention, 1939, p. 248.
310. Florence Powell Harris, p. 177.
311. *Annual,* Southern Baptist Convention, 1939, p. 248.
312. Florence Harris, letter dated Nov. 5, 1938 in Southern Baptist archives.
313. The exact number of people who died in the flood and ensuing famine is debated. It was war time, people were fleeing for their lives and no exact count was taken. *Encyclopaedia Britannica* says that estimates run from 850,000 to 4 million – the deadliest natural disaster in recorded history. "Huang He Floods," *Encyclopaedia Britannica,* n.d. 3 March 2016 http://www.britannica.com/event/Huang-He-floods.
314. Hattaway, p. 2.
315. Christensen, p. 66.
316. Christensen, p. 67.
317. *Through Fire,* p. 81.
318. Phone call with Cita Strunk on 9/21/2016.
319. Christensen, p. 66.
320. Florence Powell Harris, pp. 183 – 184.
321. Christensen, p. 226.
322. Florence Powell Harris, pp. 187-188.
323. Christensen, p. 66.
324. Adams, quoting Florence Powell Harris, p. 100.
325. *Annual,* Southern Baptist Convention, 1940, p. 226.
326. *Annual,* Southern Baptist Convention, 1941, p. 245.
327. "H. M. Harris Dies," *Foreign Mission News,* Richmond; Foreign Mission Board, Aug. 22, 1961, p. 30.
328. Gillespie, p. 17.
329. Gillespie, p. 28.
330. Identification of individuals in the photo is through the cooperation of the Nichols and Gillespie brothers (in the photo), Cita Harris Strunk, and Jim Berwick of the Archives Department of the SBC.
331. Letter by Hendon Harris Sr. to International Missions Board, dated 4/20/1941, International Missions Board of the Southern Baptist Convention Missionary Correspondence Files, AR 551-2, Box 026.
332. Florence Powell Harris, p. 186.

333. *Annual,* Southern Baptist Convention, 1940, p. 226.
334. Florence Powell Harris, p. 187.
335. Ibid.
336. Florence Powell Harris, p. 188.
337. Gillespie, p. 29.
338. Hendon Harris letter to Rankin, January 1, 1940.
339. Maddry letter to Hendon Harris, 4/22/1940.
340. Florence Powell Harris, p. 191.
341. Buford Nichols, *The Commission,* January 1941, p. 12.
342. Gillespie, p. 38.
343. Gillespie, p. 41
344. Christensen, p. 3.
345. Christensen, p. 13.
346. Christensen, p. 25.
347. Florence Powell Harris, p. 192-193, and discussion with Cita Harris Strunk by phone on 14 March, 2016. This story is given with Cita's permission.
348. Letter by Hendon Harris to International Missions Board, dated 12/25/1940, International Missions Board of the Southern Baptist Convention Missionary Correspondence Files, AR 551-2, Box 026.
349. *Annual,* Southern Baptist Convention, 1942, p. 243.
350. Letter from C.E. Maddry to H. M. Harris, dated Nov. 18, 1940.
351. Letter by Hendon Harris to International Missions Board, dated 12/25/1940, International Missions Board of the Southern Baptist Convention Missionary Correspondence Files, AR 551-2, Box 026.
352. Ibid.
353. Florence Powell Harris, p. 198.
354. Letter by Hendon Harris to International Missions Board, dated 4/20/1941, International Missions Board of the Southern Baptist Convention Missionary Correspondence Files, AR 551-2, Box 026.
355. "Hendon Mason Harris," #2606, Operation Baptist Biography Data Form for Living Person.
356. Letter by Hendon Harris to International Missions Board, dated 4/20/1941, International Missions Board of the Southern Baptist Convention Missionary Correspondence Files, AR 551-2, Box 026.
357. Hendon Harris, letter to Maddry, Dec. 9, 1941.
358. *Annual,* Southern Baptist Convention, 1942, p. 244.
359. Florence Powell Harris, p. 198.
360. Maddry letter to H. M. Harris, July 16, 1941
361. *Annual,* Southern Baptist Convention, 1942, p. 245.
362. Christensen, p. 73.
363. Adams, pp. 121-122.

364. "Dr. Charles Edward Maddry," First Baptist Church of Hillsborough, n.d. web 22 Mar 2016 http.//www.fbchillsborough.org/charles-e-maddry.
365. Charles Maddry, p. 92.
366. Russell, p. 4.
367. "Our Missionaries in the War Zones," *The Commission,* January 1942, pp. 41-49.
368. "International Mission Board Timeline," n.d. Web 11 March 2016 http://www.archives.imb.org/images/IMB%20Timeline.pdf.
369. Ibid.
370. Gillespie, p. 65.
371. Jan Jarboe Russell, *The Train to Crystal City: FDR's Secret Prisoner Exchange Program and America's Only Family Internment Camp During World War II,* New York: Scribner, reprint edition, 2016.
372. Florence Powell Harris, p. 208.
373. Florence Powell Harris, p. 215.
374. Ibid.
375. Christensen, p. 214.
376. Florence Powell Harris, p. 216.
377. Christensen, p. 219.
378. Florence Powell Harris, pp. 218-219.
379. *Journey into China,* p. 132.
380. Florence Powell Harris, p. 219.
381. Florence Powell Harris, p. 221.
382. Florence Powell Harris, p. 222.
383. Ibid.
384. Christensen, p. 230.
385. Christensen, p. 225.
386. Christensen, p. 233.
387. *Through the Fire,* p. 15.
388. Christensen, p. 231.
389. Christensen, p. 232.
390. Christensen, p. 233.
391. Christensen, p. 219.
392. Ibid.
393. Florence Powell Harris, p. 227.
394. Florence Powell Harris, p. 228.
395. Edward McDonald, *Morrison Heights Baptist Church: From the Beginning,* Book Surge Publishing, 2009, pp. 4, 17, 19, 23.
396. Adams, p. 113.
397. "Christians Now Outnumber Communists in China," 29 Dec 2014 web http://www.breibart.com/national-security/2014/12/29/christians-now-outnumber-communists-in china/.

398. Hattaway, p. 1.
399. David Aikman, *Jesus in Beijing,* Washington, DC: Regnery Publishing, 2003 pp. 67-68.
400. Aikman, p. 78.
401. Aikman, p. 87.
402. Hendon Harris, "Indigenous Churches in China," pp. 245-247.
403. Keay, pp. 38, 39.
404. Hendon Harris, "Indigenous Churches in China," p. 137.
405. Hendon Harris Jr. *The Asiatic Fathers of America: The Chinese Discovery and Colonization of Ancient America,* Book 1, Taipei: Wen Ho Printing Co, Ltd, 1973, p. 11.
406. Lawrence Harris, notes on proposed manuscript of mine that he read, January 2010.
407. Recorded verbal family history by Marjorie Weaver Harris.
408. Phone conversation with Cita Strunk, 9/21/2016.
409. John 14:6, Holy Bible, NKJV
410. E-mail correspondence from Roger Voegtlin, 8/4/2016.
411. Album belonging to Harris Family.
412. Ibid.
413. Shu Guang Zhang, *Deterrence and Strategic Culture: Chinese-American Confrontations, 1949-1958,* Ithaca, NY: Cornell U. Press, 1993, p. 52.
414. Zhang, p. 54.
415. Zhang, p. 57.
416. Zhang, p. 58.
417. Zhang, p. 55.
418. Ibid.
419. Hendon M. Harris, *The American Idol*, 1961, p. 3.
420. Hendon M. Harris, *The Asiatic Fathers of America*, Bk. 1, p. 167.
421. "Other News to Note Deaths – Spencer Moosa" 22 May 1990 web 17 July 2016 www.http://articles.orlandosentinel.com/1990-05-22/news/9005220211_1_moosa-chiang-kai-shek-spencer.
422. Scrapbook belonging to Harris family.
423. Yang Yang, *The Sage in the Cathedral of Books: The Distinguished Chinese American Library Professional Dr. Hwa-Wei Lee,* tr. by Ying Zhang, Athens, Ohio: Ohio University Special Publications, p. 19, 83.
424. *Annual,* Southern Baptist Convention, 1951, p. 159.
425. *Annual,* Southern Baptist Convention, 1951, p. 160.
426. Hendon M. Harris Jr., *Laughter and Tears,* Berne, IN: Light and Hope Publishing, 1954, p. 39.
427. Ibid, p. 66.
428. Hendon M. Harris Jr., *Poems for Grown Up Children*, Berne, IN: Light and Hope Publications, ca. 1953.
429. Hendon M. Harris Jr., *Laughter and Tears.*

430. “The Gospel Work in the Mountain of Taiwan,” Brochure by the Hundred Nations Crusade, Taitung, Taiwan.
431. Harris, *The Asiatic Fathers of America,* Bk. 1, p. 167.
432. Translation to English by Naomi Zhang of Chinese inscription in Harris Album.
433. “World Freedom Day in Taiwan,” https://anydayguide.com/calendar/39.
434. Robert L. Alderman, Letter to Charlotte Rees, 4/23/2016.
435. Hendon and Marjorie Harris prayer letter dated December 1966.
436. Hendon and Marjorie Harris prayer letter dated April 1967.
437. Hendon Harris Jr., Prayer Letter of Hundred Nations Crusade, May 1967.
438. Hendon and Marjorie Harris prayer letter dated June 1967
439. Hendon and Marjorie Harris prayer letter dated March 1970
440. Hendon and Marjorie Harris prayer letter dated August 1971
441. California, Divorce Index, 1966-1984.
442. Hendon Harris Jr, *The Asiatic Fathers of America,* Book 1, pp. 16-18.
443. *Holy Bible,* New International Version, Ephesians 2:8 - 9.
444. *Holy Bible:* I Corinthians 15:3.

Additional Resources

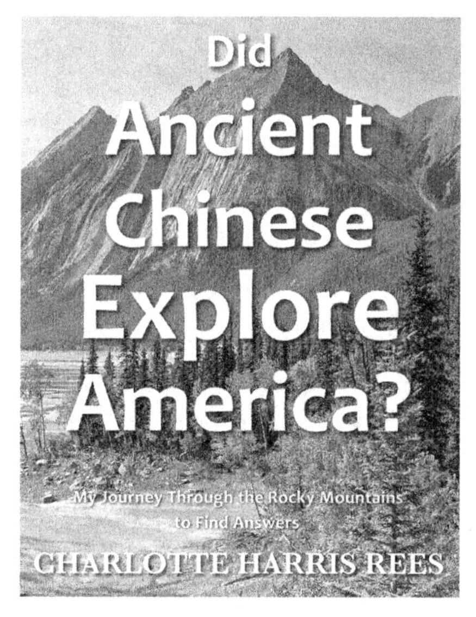

***Did Ancient Chinese Explore America? My Journey Through the Rocky Mountains to Find Answers*—** tests the hypothesis that ancient Chinese geographical descriptions of the "Eastern Mountains" actually referenced locations in North America. In this travelogue Rees candidly shares her initial doubts then her search and discoveries as she follows an 1100 mile trek through the Rocky Mountains. She weaves together history, subtle humor, academic studies, and many photographs to tell a compelling story.

***Secret Maps of the Ancient World*—** brings together an abundance of worldwide academic evidence that Asians reached the Americas at very early dates. It also compares peoples on both sides of the Pacific, discusses DNA, and relates how Asians could have traveled to the Americas by sea. The capstone of the book is an ancient Asian map shown in museums around the world, which until recently was thought to be part imaginary. The thesis of the book is that secrets on those maps were hidden for years in plain sight.

***The Asiatic Fathers of America: Chinese Discovery and Colonization of Ancient America*—** emphasizes the many early Chinese writings that discussed Fu Sang, a beautiful land to the East of China. However, over thousands of years most forgot where Fu Sang was. Then in 1972 Dr. Hendon Harris Jr. (Charlotte Harris Rees's father) discovered an antique Asian map which shows Fu Sang on the American coastline. Harris discusses early Chinese literary descriptions of the Grand Canyon and other geographical features of North America. He compares human characteristics. This book is Charlotte's abridgment and edit of her father's original tome of almost 800 pages.

New World Secrets on Ancient Asian Maps—
explores historical settings and distinctive styles of various old Asian maps and relates interesting curiosities found on them—including ancient love notes, state secrets, and internationally volatile data. These secrets could change your understanding of world history.

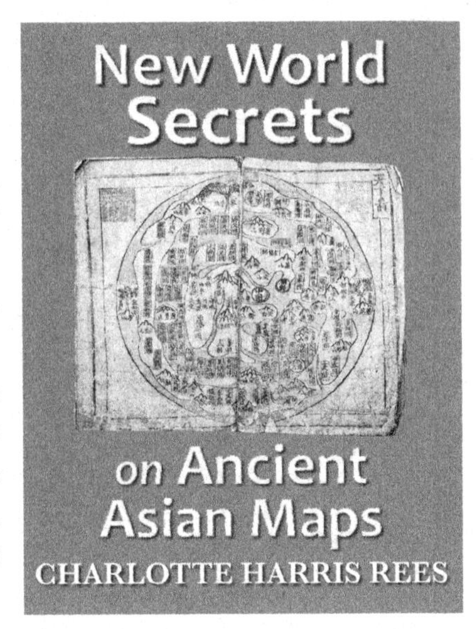

Initially a skeptic of the theory of early arrival of Chinese to the Americas, the author's initial research was to find the meaning of the maps in the prized Dr. Hendon M. Harris Jr. Collection. Her research took her beyond that to various related maps and what they reveal about Chinese knowledge of the New World at early dates. This book presents additional text, clearer maps, and more illustrations than it did under its previous title: *Chinese Sailed to America Before Columbus.*

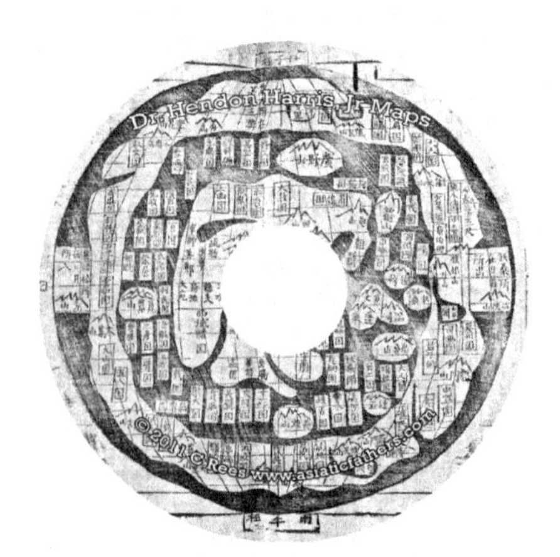

Dr. Hendon M. Harris Jr. Map Collection on CD—
contains over 100 high definition photos showing all the maps in the Harris collection - some both front and back. Available only through www.AsiaticFathers.com

A Short History of Religious Liberty—
doctoral thesis of Hendon M. Harris Jr. This thesis recounts views on religious liberty throughout European and US history.

Indigenous Churches in China—
doctoral thesis of Hendon M. Harris Sr. (available in portable document format.)

www.AsiaticFathers.com

CPSIA information can be obtained
at www.ICGtesting.com
Printed in the USA
FSOW03n2301251116
27692FS